ENGINEERING OPTICS WITH MATLAB®

Second Edition

ENGINEERING OPTICS WITH MATLAB®

Second Edition

Ting-Chung Poon
Virginia Tech, USA

Taegeun Kim
Sejong University, South Korea

NEW JERSEY · LONDON · SINGAPORE · BEIJING · SHANGHAI · HONG KONG · TAIPEI · CHENNAI · TOKYO

Published by

World Scientific Publishing Co. Pte. Ltd.

5 Toh Tuck Link, Singapore 596224

USA office: 27 Warren Street, Suite 401-402, Hackensack, NJ 07601

UK office: 57 Shelton Street, Covent Garden, London WC2H 9HE

Library of Congress Cataloging-in-Publication Data

Names: Poon, Ting-Chung, author. | Kim, Taegeun, author.
Title: Engineering optics with MATLAB / by Ting-Chung Poon (Virginia Tech, USA),
    Taegeun Kim (Sejong University, South Korea).
Description: 2nd edition. | [Hackensack] New Jersey : World Scientific, 2017. |
    Includes index.
Identifiers: LCCN 2017016687| ISBN 9789813100008 (hc : alk. paper) |
    ISBN 9789813100015 (pbk : alk. paper)
Subjects: LCSH: Optics--Data processing. | Optical engineering. |
    Fourier transform optics. | Acoustooptics. | Electrooptics. | MATLAB.
Classification: LCC QC355.3 .P66 2017 | DDC 621.360285/53--dc23
LC record available at https://lccn.loc.gov/2017016687

British Library Cataloguing-in-Publication Data
A catalogue record for this book is available from the British Library.

Desk Editor: Herbert Moses

Typeset by Stallion Press
Email: enquiries@stallionpress.com

Printed in Singapore

To the memory of my parents.

- Ting-Chung Poon

To Sang Min Lee, Chanyun Kim, and Siyeon Kim.

- Taegeun Kim

Preface to the Second Edition

This edition of *Engineering Optics with MATLAB®* provides more in-depth discussions and examples in various chapters. The underlying philosophy of this edition is the same as that in the first edition, which serves two purposes. The first introduces some traditional topics such as matrix formalism of geometrical optics, wave propagation and diffraction, and some fundamental background on Fourier optics. The second presents the essentials of acousto-optics and electro-optics, and provides the students with experience in modeling the theory and applications using a commonly used software tool *MATLAB®*.

Preface to the First Edition

This book serves two purposes: The first is to introduce the readers to some traditional topics such as the matrix formalism of geometrical optics, wave propagation and diffraction, and some fundamental background on Fourier optics. The second is to introduce the essentials of acousto-optics and electro-optics, and to provide the students with experience in modeling the theory and applications using MATLAB®, a commonly used software tool. This book is based on the authors' own in-class lectures as well as research in the area.

The key features of the book are as follows. Treatment of each topic begins from the first principles. For example, geometrical optics starts from Fermat's principle, while acousto-optics and electro-optics start from

Maxwell equations. MATLAB examples are presented throughout the book, including programs for such important topics as diffraction of Gaussian beams, split-step beam propagation method for beam propagation in inhomogeneous as well as Kerr media, and numerical calculation of up to 10-coupled differential equations in acousto-optics. Finally, we cover acousto-optics with emphasis on modern applications such as spatial filtering and heterodyning.

The book can be used as a general text book for Optics/Optical Engineering classes as well as acousto-optics and electro-optics classes for advanced students. It is our hope that this book will stimulate the readers' general interest in optics as well as provide them with an essential background in acousto-optics and electro-optics. The book is geared towards a senior/first-year graduate level audience in engineering and physics. This is suitable for a two-semester course. The book may also be useful for scientists and engineers who wish to learn about the basics of beam propagation in inhomogeneous media, acousto-optics and electro-optics.

Ting-Chung Poon (TCP) would like to thank his wife Eliza and his children Christina and Justine for their encouragement, patience and love. In addition, TCP would like to thank Justine Poon for typing parts of the manuscript, Bill Davis for help with the proper use of the word processing software, Ahmad Safaai-Jazi and Partha Banerjee for help with better understanding of the physics of fiber optics and nonlinear optics, respectively, and last, but not least, Monish Chatterjee for reading the manuscript and providing comments and suggestions for improvements.

Taegeun Kim would like to thank his wife Sang Min Lee and his parents Pyung Kwang Kim and Ae Sook Park for their encouragement, endless support and love.

About the Authors

Ting-Chung Poon is a Professor of Electrical and Computer Engineering at Virginia Tech, Virginia, USA. His current research interests include 3-D image processing, and optical scanning holography. Dr. Poon is the author of the monograph *Optical Scanning Holography with MATLAB* (Springer, 2007), and is the co-author of the textbooks *Introduction to Modern Digital Holography with MATLAB* (Cambridge University Press, 2014), *Engineering Optics with MATLAB* (World Scientific, 2006), *Contemporary Optical Image Processing with MATLAB* (Elsevier, 2001), and *Principles of Applied Optics* (McGraw-Hill, 1991). He is also the Editor of the book *Digital Holography and Three-Dimensional Display* (Springer, 2006). Dr. Poon served as Topical Editor/Division Editor of *Applied Optics* from 2004 to 2014. He was also Associate Editor of the *IEEE Transactions on Industrial Informatics*. Currently Dr. Poon is Associate Editor-in-Chief of *Chinese Optics Letters*. Dr. Poon is the founding Chair of the Optical Society (OSA) topical meeting Digital holography and 3-D imaging (2007). Dr. Poon is a Fellow of the Institute of Electrical and Electronics Engineers (IEEE), the Institute of Physics (IOP), the OSA, and the International Society for Optics and Photonics (SPIE). He received the 2016 Dennis Gabor Award of the SPIE for "his pioneering contributions to optical scanning holography (OSH), which has contributed significantly to the development of novel digital holography and 3-D imaging."

Taegeun Kim is a Professor of Electrical Engineering at Sejong University, Seoul, Korea. His current research interests include holography and 3-D imaging systems. Dr. Kim is the co-author of the textbook *Engineering Optics with MATLAB* (World Scientific, 2006). He served as a Topical

Editor of the *Journal of Optical Society of Korea* from 2011 to 2014, and he has been a Topical Editor of *Applied Optics* since 2015. Dr. Kim is a member of the Optical Society of Korea (OSK), and the Optical Society of America (OSA). He is also a Senior Member of the International Society for Optics and Photonics (SPIE).

Contents

Chapter 1

Geometrical Optics

When we consider optics, the first thing that comes to our minds is probably light. Light has a dual nature: light is particles (called photons) and light is waves. When a particle moves, it processes momentum, p. And when a wave propagates, it oscillates with a wavelength, λ. Indeed, the momentum and the wavelength is given by the *de Broglie relation*

$$\lambda = \frac{h}{p},$$

where $h \approx 6.2 \times 10^{-34}$ Joule-second is Planck's constant. Hence from the relation, we can state that every particle is a wave as well.

Each particle or photon is specified precisely by the frequency v and has an energy E given by

$$E = hv.$$

If the particle is traveling in free space or in vacuum, the frequency is $v = c / \lambda$, where c is a constant approximately given by $3 \times 10^8 \, m / s$. The speed of light in a transparent linear, homogeneous and isotropic material, which we term u, is again a constant but less than c. This constant is a physical characteristic or signature of the material. The ratio c / u is called the *refractive index*, n, of the material.

In *geometrical optics*, we treat light as particles and the trajectory of these particles follows along paths that we call *rays*. We can describe an optical system consisting of elements such as mirrors and lenses by tracing the rays through the system.

Geometrical optics is a special case of *wave* or *physical* optics, which will be mainly our focus through the rest of this chapter. Indeed, by taking the limit in which the wavelength of light approaches zero in wave optics, we recover geometrical optics. In this limit, diffraction and the wave nature of light is absent.

1.1 Fermat's Principle

Geometrical optics starts from *Fermat's Principle*. In fact, Fermat's Principle is a concise statement that contains all the physical laws, such as the *law of reflection* and *the law of refraction*, in geometrical optics. Fermat's principle states that the path of a light ray follows is an extremum in comparison with the nearby paths. The extremum may be a minimum, a maximum, or stationary with respect to variations in the ray path. However, it is usually a minimum. In Figs. 1.1a and b, we show the situations for a maximum and being stationary, respectively (the situation for a minimum will be discussed when we talk about the laws of reflection and refraction in the next section).

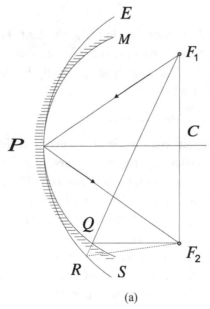

(a)

Fig. 1.1. (a) Spherical mirror.

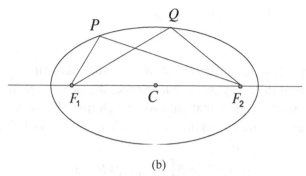

(b)

Fig. 1.1. (b) Elliptical mirror.

In Fig. 1.1a, we inspect reflection from a spherical mirror (M). Point C represents the center of curvature of the spherical mirror and the distance CP is the radius of curvature of the mirror. We also construct an ellipse (E) passing through P with foci F_1 and F_2. Now consider light starting from F_1 and reaching F_2 upon reflection from P. The ray's path length is $F_1P + PF_2$. Now consider another path length $F_1Q + F_2Q$, where Q is another point on the spherical mirror. Since $QR + RF_2 > QF_2$, where R is the point on the ellipse when we extend line F_1Q to intercept the ellipse, $F_1Q + QR + RF_2 > F_1Q + QF_2$. So $F_1R + RF_2 > F_1Q + QF_2$. Now from the property of an ellipse, we know $F_1P + PF_2 = F_1R + RF_2$. Hence $F_1P + PF_2 > F_1Q + QF_2$, i.e., the ray of an actual ray path is always longer than a neighboring path and this case represents that the extremum is a maximum. Figure 1.1b shows an elliptical mirror on the top half of the ellipse with foci F_1 and F_2. We can see that any ray starting from one of the foci, say F_1, after being reflected from any point (P or Q) on the mirror will pass through the other focus, F2. From the property of an ellipse, $F_1P + PF_2 = F_1Q + QF_2$. Hence all rays after reflection have the same path length and this is the case that that the extremum is stationary.

We now give a mathematical description of Fermat's principle. Let $n(x, y, z)$ represent a position-dependent refractive index along a path C between end points A and B, as shown in Fig. 1.1c. We define the *optical path length (OPL)* as

$$OPL = \int_C n(x,y,z)ds,$$ (1.1-1)

where ds represents an infinitesimal arc length. According to Fermat's principle, out of the many paths that connect the two end points A and B, the light ray would follow that path for which the OPL between the two points is an extremum, i.e., in the context of calculus of variations,

$$\delta(OPL) = \delta\int_C n(x,y,z)ds = 0$$ (1.1-2)

in which δ represents a small variation. In other words, a ray of light will travel along a medium in such a way that the total OPL assumes an extremum. As an extremum means that the rate of change is zero, Eq.(1.1-2) explicitly means that

$$\frac{\partial}{\partial x}\int nds + \frac{\partial}{\partial y}\int nds + \frac{\partial}{\partial z}\int nds = 0.$$ (1.1-3)

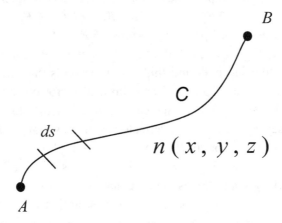

Fig. 1.1 (c) A ray of light traversing a path C between end points A and B.

Now since the ray propagates with the velocity $u = c / n$ along the path,

$$nds = \frac{c}{u} ds = cdt, \qquad (1.1\text{-}4)$$

where dt is the differential time needed to travel the distance ds along the path. We substitute Eq. (1.1-4) into Eq. (1.1-2) to get

$$\delta \int_C nds = c\delta \int_C dt = 0. \qquad (1.1\text{-}5)$$

As mentioned before, the extremum is usually a minimum, we can, therefore, restate Fermat's principle as a *principle of least time*. In a *homogeneous medium*, i.e., in a medium with a constant refractive index, the ray path is a straight line as the smallest *OPL* between the two end points is along a straight line which assumes the shortest time for the ray to travel.

1.2 Reflection and Refraction

When a ray of light is incident on the interface separating two different optical media characterized by n_1 and n_2, as shown in Fig. 1.2, it is well known that part of the light is reflected back into the first medium, while the rest of the light is refracted as it enters the second medium. The directions taken by these rays are described by the laws of reflection and refraction, which can be derived from Fermat's principle.

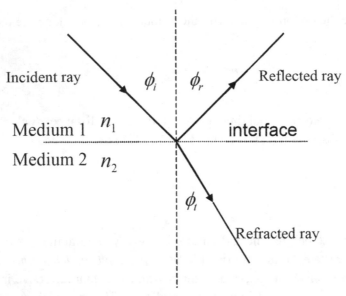

Fig. 1.2 Reflected and refracted rays for light incident at the interface of two media.

In what follows, we demonstrate the use of the principle of least time to derive the law of reflection. Consider a reflecting surface as shown in Fig. 1.3. Light from point A is reflected from the reflecting surface to point B, forming the angle of incidence ϕ_i and the angle of reflection ϕ_r, measured from the normal to the surface. The time required for the ray of light to travel the path $AO + OB$ is given by $t = (AO + OB)/u$, where u is the velocity of light in the medium containing the points AOB. The medium is considered isotropic and homogeneous. From the geometry, we find

$$t(z) = \frac{1}{u}([h_1^2 + (d-z)^2]^{1/2} + [h_2^2 + z^2]^{1/2}).$$

(1.2-1)

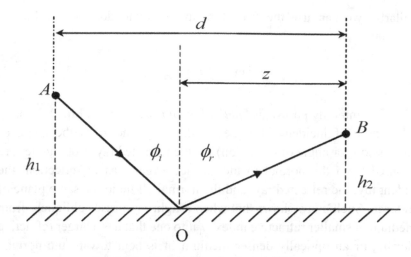

Fig. 1.3 Incident (AO) and reflected (OB) rays.

According to the least time principle, light will find a path that extremizes $t(z)$ with respect to variations in z. We thus set $dt(z)/dz$ to get

$$\frac{d-z}{[h_1^2 + (d-z)^2]^{1/2}} = \frac{z}{[h_2^2 + z^2]^{1/2}} \qquad (1.2\text{-}2)$$

or

$$\sin\phi_i = \sin\phi_r \qquad (1.2\text{-}3)$$

so that

$$\phi_i = \phi_r, \qquad (1.2\text{-}4)$$

which is the law of reflection. We can readily check that the second derivative of $t(z)$ is positive at critical point z_1, where $dt(z_1)/dz = 0$ so that the result obtained corresponds to the least time principle. In addition, Fermat's principle also demands that the incident ray, the reflected ray and the normal all be in the same plane, called the *plane of incidence*.

Similarly, we can use the least time principle to derived the law of refraction

$$n_1 \sin \phi_i = n_2 \sin \phi_t \; , \qquad (1.2\text{-}5)$$

which is commonly known as *Snell's law of refraction*. In Eq. (1.2-5), ϕ_i is the angle of incidence for the incident ray and ϕ_t is the angle of transmission (or angle of refraction) for the refracted ray. Both angles are measured from the normal to the surface. Again, as in reflection, the incident ray, the refracted ray, and the normal all lie in the same plane of incidence. Snell's law shows that when a light ray passes obliquely from a medium of smaller refractive index n_1 into one that has a larger refractive index n_2, or an optically denser medium, it is bent toward the normal. Conversely, if the ray of light travels into a medium with a lower refractive index, it is bent away from the normal. For the latter case, it is possible to visualize a situation where the refracted ray is bent away from the normal by exactly $90°$. Under this situation, the angle of incidence is called the *critical angle* ϕ_c, given by

$$\sin \phi_c = n_2 \, / \, n_1 . \qquad (1.2\text{-}6)$$

When the incident angle is greater than the critical angle, the ray originating in medium 1 is totally reflected back into medium 1. This phenomenon is called *total internal reflection*. The optical fiber uses this principle of total reflection to guide light, and the mirage on a hot summer day is a phenomenon due to the same principle.

1.3 Ray Propagation in an Inhomogeneous Medium: Ray Equation

In the last Section, we have discussed refraction between two media with different refractive indices, possessing a discrete inhomogeneity in the simplest case. For a general inhomogeneous medium, i.e., $n(x,y,z)$, it is instructive to have an equation that can describe the trajectory of a ray. Such an equation is known as the *ray equation*. The ray equation is analogous to the equations of motion for particles and for rigid bodies in classical mechanics. The equations of motion can be derived from *Newtonian mechanics* based on Newton's laws. Alternatively, the equations of motion can be derived directly from *Hamilton's principle of least action*. Indeed Fermat's principle in optics and Hamilton's principle of least action in classical mechanics are analogous. In what follows, we describe Hamilton's principle so as to formulate the so called *Lagrange's equations* in mechanics. We then re-formulate Lagrange's equations for optics to derive the ray equation.

Hamilton's principle states that the trajectory of a particle between times t_1 and t_2 is such that the variation of the line integral for fixed t_1 and t_2 is zero, i.e.,

$$\delta \int_{t_1}^{t_2} L(q_k, \dot{q}_k, t)dt = 0, \qquad (1.3\text{-}1)$$

where $L = T - V$ is known as the *Lagrangian function* with T being the kinetic energy and V the potential energy of the particle. The q_k's are called *generalized coordinates* with $k = 1, 2, 3, ...n$. Also, $\dot{q}_k = dq_k / dt$. Generalized coordinates are any collection of independent coordinates q_k (not connected by any equations of constraint) that are sufficient to specify uniquely the motion. The number n of generalized coordinates is the number of *degrees of freedom*. For example, a simple pendulum has one degree of freedom, i.e., $q_k = q_1 = \phi$, where ϕ is the angle the pendulum makes with the vertical. Now if the simple pendulum is complicated such that the string holding the bob is elastic. There will be two generalized coordinates, $q_k = q_1 = \phi$, and $q_k = q_2 = x$, where x is the length of the string. As another example, let us consider a particle constrained to move along the surface of a sphere with radius R. The coordinates (x, y, z) do not constitute an independent set as they are connected by the equation of constraint $x^2 + y^2 + z^2 = R^2$. The particle has only two degrees of freedom and two independent coordinates are needed to specify its position on the sphere uniquely. These coordinates could be taken as latitude and longitude or we could choose angles θ and ϕ from spherical coordinates as our generalized coordinates.

Now, if the force field F is conservative, i.e., $\nabla \times F = 0$, the total energy $E = T + V$ is a constant during the motion, and Hamilton's principle leads to the following equations of motion of the particle called *Lagrange's equations*:

$$\frac{d}{dt}\left(\frac{\partial L}{\partial \dot{q}_K}\right) = \frac{\partial L}{\partial q_K}. \tag{1.3-2}$$

As a simple example illustrating the use of Lagrange's equations, let us consider a particle with mass m having kinetic energy $T = m|\dot{r}|^2/2$ under potential energy $V(x, y, z)$, where $r(x, y, z) = x(t)a_x + y(t)a_y + z(t)a_z$ is the position vector with a_x, a_y, and a_z being the unit vector along the x, y, and z direction, respectively. According to Newton's second law,

$$F = m\ddot{r}, \tag{1.3-3}$$

where $\ddot{r}$ is the second derivative of r with respect to t. As usual the force is given by the negative gradient of the potential, i.e., $F = -\nabla V$. Hence, we have the vector equation of motion for the particle

$$m\ddot{r} = -\nabla V \tag{1.3-4}$$

according to Newtonian mechanics.

Now from the Lagrange's equations, we identify $L = T - V = \frac{1}{2}m|\dot{r}|^2 - V$. Considering, $q_1 = x$, we have

$$\frac{d}{dt}\left(\frac{\partial L}{\partial \dot{x}}\right) = m\ddot{x} \text{ and } \frac{\partial L}{\partial x} = -\frac{\partial V}{\partial x}. \tag{1.3-5}$$

From Eq. (1.3-2) and using the above results, we have

$$m\ddot{x} = -\frac{\partial V}{\partial x}, \tag{1.3-6}$$

and similarly for the y and z components as $q_2 = y$ and $q_3 = z$. Therefore, we come up with Eq. (1.3-4), which is directly from Newtonian mechanics. Hence, we see that Newton's equations can be derived from Lagrange's equations and in fact, the two sets of equations are equally fundamental. However, the Lagrangian formalism has certain advantages over the conventional Newtonian laws in that the physics problem has been transformed into a purely mathematical problem. We just need to find T and V for the system and the rest is just mathematical consideration through the use of Lagrange's equations. In addition, there is no need to consider any vector equations as in Newtonian mechanics as Lagrange's equations are scalar quantities. As it turns out, Lagrange's equations are much better adapted for treating complex systems such as in the areas of quantum mechanics and general relativity.

===

The Simple Pendulum Example

With reference to the figure in the Example, we have a simple pendulum problem: a point mass of m attached to the end of a rod of negligible mass; the motion being confined in a vertical plane.

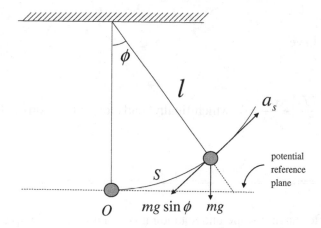

The simple pendulum example.

The arc length s is measured from the equilibrium position O. a_s is the unit vector tangential to the arc. From Newton's second law, we have

$$F = m\ddot{s}a_s = -mg\sin\phi\, a_s \,,$$

where mg is the force of gravity. The minus sign denotes the restoring force which tries to bring the point mass back to equilibrium. Now $s = l\phi$, where l is the length of the pendulum. The above equation then becomes

$$l\ddot{\phi} = -g\sin\phi \ \text{ or } \ \ddot{\phi} + (g/l)\sin\phi = 0.$$

Using the Lagrangian formalism, we first find T and V for the system under consideration:

$$T = mu^2/2 = m(l\dot{\phi})^2/2; \quad V = mg(l - l\cos\phi),$$

and $L = T - V$. We can then find

$$\frac{\partial L}{\partial \dot{\phi}} = ml^2\dot{\phi}\,; \ \ \frac{\partial L}{\partial \phi} = -mgl\sin\phi\,.$$

Hence, we have

$$\frac{d}{dt}\left(\frac{\partial L}{\partial \dot{\phi}}\right) - \frac{\partial L}{\partial \phi} = 0, \text{ which finally leads to } \ddot{\phi} + (g/l)\sin\phi = 0.$$

==

After having some understanding of Hamilton's principle, and the use of Lagrange's equations to obtain the equations of motion of a particle, we now formulate Lagrange's equations in optics. Again, the particles of concern in optics are photons. Starting from Fermat's principle as given by Eq. (1.1-2),

$$\delta \int_C n(x, y, z) ds = 0.$$ (1.3-7)

We write the arc length *ds* along the path of the ray as

$$ds^2 = dx^2 + dy^2 + dz^2$$ (1.3-8)

with reference to Fig. 1.4, where for brevity, we have only shown the 2-D (i.e., $x - z$) version of the configuration. Defining $x' = dx / dz$ and $y' = dy / dz$, we can write Eq. (1.3-8) as

$$ds = dz \sqrt{1 + (x')^2 + (y')^2}.$$ (1.3-9)

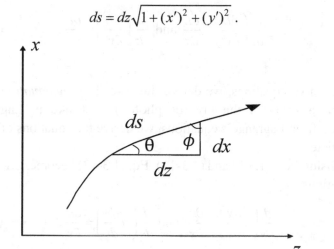

Fig. 1.4 The path of a ray in a continuous Inhomogeneous medium.

Substituting Eq. (1.3-9) into Eq. (1.3-7), we have

$$\delta \int_C n(x, y, z) \sqrt{1 + (x')^2 + (y')^2} \, dz = 0.$$ (1.3-10)

By comparing this equation with Eq. (1.3-1), we can define the so-called *optical Lagrangian* as

$$L(x,y,x',y',z) = n(x,y,z)\sqrt{1+(x')^2+(y')^2}. \qquad (1.3\text{-}11)$$

We can see that Hamilton's principle is based on minimizing functions of time, whereas Fermat's principle minimizes a function of length, z, as we have assumed z to play the same role as t in Lagrangian mechanics, where we have chosen the z-direction as the direction along which the rays are propagating. Now that we have established the optical Lagrangian, we can immediately write down the following Lagrange's equations in optics by referring to Eq. (1.3-2):

$$\frac{d}{dz}\left(\frac{\partial L}{\partial x'}\right)=\frac{\partial L}{\partial x} \text{ and } \frac{d}{dz}\left(\frac{\partial L}{\partial y'}\right)=\frac{\partial L}{\partial y}. \qquad (1.3\text{-}12)$$

From these two equations, we derive the so-called *ray equation*, which tracks the position of the ray (or photon); just like in Lagrangian mechanics, from Lagrange's equations, we derive the equations of motion for a particle.

Using Eqs. (1.3-9) and (1.3-11), Eq. (1.3-12) becomes, after some manipulations,

$$\frac{d}{ds}\left(n\frac{dx}{ds}\right)=\frac{\partial n}{\partial x} \text{ and } \frac{d}{ds}\left(n\frac{dy}{ds}\right)=\frac{\partial n}{\partial y}. \qquad (1.3\text{-}13)$$

The objective is of course, for a given n, to find $x(s)$ and $y(s)$ by solving the above equations. It is important to point out that the two equations above are sufficient to determine the ray trajectory. This indicates that the z-component of the ray equation is really redundant. Indeed the corresponding equation for z, given below, can be derived from the equations for x and y:

$$\frac{d}{ds}\left(n\frac{dz}{ds}\right)=\frac{\partial n}{\partial z}\ .\qquad\qquad(1.3\text{-}14)$$

Now the desired *ray equation in vectorial form* is obtained by combining Eqs. (1.3-13) and (1.3-14)

$$\frac{d}{ds}\left(n\frac{d\mathbf{r}}{ds}\right)=\nabla n,\qquad\qquad(1.3\text{-}15)$$

where once again $\mathbf{r}(s)$ is a position vector which represents the position of any point on the ray.

Example 1.1 Homogeneous Medium

For $n(x,y,z)=$ constant, Eq. (1.3-15) becomes

$$\frac{d^2\mathbf{r}}{ds^2}=0,\qquad\qquad(1.3\text{-}16)$$

which has solutions

$$\mathbf{r}=\mathbf{a}s+\mathbf{b},\qquad\qquad(1.3\text{-}17)$$

where $\mathbf{a}$ and $\mathbf{b}$ are some constant vectors determined from the initial conditions, and Eq. (1.3-17) is clearly a straight line equation for the ray path in a homogenous medium. The situation is shown in Fig. 1.5.

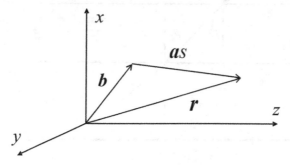

Fig. 1.5 Ray propagating along a straight line in homogenous medium.

Example 1.2 Law of refraction derived from the ray equation

We consider a 2-D situation involving x and z coordinates where n is a function of x only. The medium consists of a set of thin slices of media of different refractive indices as shown in Fig. 1.6. Since we are interested in how the ray travels along z, we can use Eq. (1.3-14), which becomes

$$\frac{d}{ds}\left(n\frac{dz}{ds}\right) = 0 \,,$$

or

$$n\frac{dz}{ds} = \text{constant.}$$

Since $dz / ds = \cos\theta = \sin\phi$ (see Fig. 1.4), the above equation can written as

$$n\sin\phi = \text{ constant} \,,$$

which holds true throughout the ray trajectory. Hence we have derived the law of refraction, or Snell's law, i.e., $n_1 \sin\phi_1 = n_2 \sin\phi_2 = n_3 \sin\phi_3$.

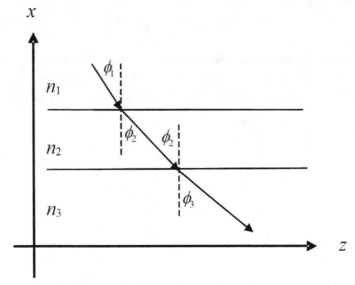

Fig. 1.6 Ray refracted along layers of discrete medium.

Example 1.3 Square-Law Medium:

$$n^2(x, y) = n_0^2 - n_2(x^2 + y^2)$$

In this example, we first consider an optical waveguide with a z- independent refractive index, i.e., $n = n(x, y)$ in general, and then find a solution to a special case of a *square-law medium* where $n^2(x, y) = n_0^2 - n_2(x^2 + y^2)$. Note that n_2 is considered small enough such that $n_0^2 \gg n_2(x^2 + y^2)$ for all practical values of x and y.

For the case that $n(x, y)$ is not a function of z, we can inspect Eq. (1.3-14) to get some insight into the problem:

$$\frac{d}{ds}\left(n \frac{dz}{ds} \right) = \frac{\partial n}{\partial z} = 0 ,$$

which means that $n \frac{dz}{ds}$ is not a function of s , i.e., it is constant along the ray path. In fact, it is not a function of any coordinates x, y and z as $s(x, y, z)$. Hence $n \frac{dz}{ds}$ is strictly a constant. We then let $n \frac{dz}{ds} = \tilde{\beta}$ and refer to Fig. 1.4 to use the fact that $\frac{dz}{ds} = \cos\theta(x, y)$, and by taking into account the y-dimension for a general situation, we arrive at an equation

$$n(x, y)\cos\theta(x, y) = \tilde{\beta}. \qquad (1.3\text{-}18)$$

The above equation is *generalized Snell's law* and it means physically that as the ray travels along a trajectory inside the waveguide, the ray would bend in such a way that the product $n(x, y)\cos\theta(x, y)$ or $n(x, y)\sin\phi(x, y)$ remains the same. Now let us find the equations so that we can solve for $x(z)$ and $y(z)$. To find $x(z)$, we can use Lagrange's equation involving x [see Eq. (1.3-12)], i.e.,

$$\frac{d}{dz}\left(\frac{\partial L}{\partial x'}\right) = \frac{\partial L}{\partial x} \qquad (1.3\text{-}19)$$

with $L = n(x,y)\sqrt{1+(x')^2 +(y')^2}$ for our current example. We can show that Eq. (1.3-19) becomes

$$\frac{d^2 x}{dz^2} = \frac{n}{\tilde{\beta}^2}\frac{\partial n}{\partial x} = \frac{1}{2\tilde{\beta}^2}\frac{\partial n^2}{\partial x}. \qquad (1.3\text{-}20)$$

Similarly, we can derive the ray equation for $y(z)$ by using the y-component of Eq. (1.3-12):

$$\frac{d^2 y}{dz^2} = \frac{n}{\tilde{\beta}^2}\frac{\partial n}{\partial y} = \frac{1}{2\tilde{\beta}^2}\frac{\partial n^2}{\partial y}. \qquad (1.3\text{-}21)$$

The above two equations are rigorous equations for media with the index of refraction independent of z.

We now consider a simple example in the square-law medium and find the ray path for propagation in x-z plane when we launch a ray from $x = x_0$ with a launching angle α with respect to the z-axis. We use Eq. (1.3-20), which becomes

$$\frac{d^2 x}{dz^2} = -\frac{n_2}{n^2(x_0)\cos^2\alpha}x(z), \qquad (1.3\text{-}22)$$

where we have used the definition that

$$\tilde{\beta} = n(x_0)\cos\theta(x_0) = n(x_0)\cos\alpha.$$

Equation (1.3-22) has a general solution of the form given

$$x(z) = A\sin\left(\frac{\sqrt{n_2}}{n(x_0)\cos\alpha}z + \phi_0\right), \qquad (1.3\text{-}23)$$

where the constants A and ϕ_0 can be determined from the initial position and slope of the ray. Note that rays with smaller launching angles α have a larger period; however, in the paraxial approximation (i.e., for small launching angles), all the ray paths have approximately the same period. These rays, which lie in the plane containing the so-called *optical axis* (z- axis), are called *meridional rays* and all other rays are called *skew rays*.

Let us now discuss a case in that the ray is launched on the y-z plane at $x = x_0$, $y = 0$ and $z = 0$ with a launching angle α with respect to the z-axis. Under these considerations, Eqs. (1.3-20) and (1.3-21) become

$$\frac{d^2 x}{dz^2} = -\frac{n_2}{\tilde{\beta}^2} x(z) \qquad (1.3\text{-}24a)$$

and

$$\frac{d^2 y}{dz^2} = -\frac{n_2}{\tilde{\beta}^2} y(z), \qquad (1.3\text{-}24b)$$

respectively, where $\tilde{\beta} = n(x, y)\cos\theta(x, y) = n(x_0, 0)\cos\alpha$.

The corresponding boundary conditions for Eqs. (1.3-24a) and (1.3-24b) are

$$x(0) = x_0, \frac{dx(0)}{dz} = 0 \qquad (1.3\text{-}25a)$$

and

$$y(0) = 0, \frac{dy(0)}{dz} = \tan\alpha. \qquad (1.3\text{-}25b)$$

The solutions of Eqs. (1.3-24a) and (1.3-24b) are

$$x(z) = x_0 \cos\left(\frac{\sqrt{n_2}}{\tilde{\beta}} z\right) \qquad (1.3\text{-}26a)$$

and

$$y(z) = \tilde{\beta}\frac{\tan\alpha}{\sqrt{n_2}}\sin\left(\frac{\sqrt{n_2}}{\tilde{\beta}}z\right), \qquad (1.3\text{-}26b)$$

respectively. In general, the two equations are used to describe skew rays. As a simple example, if $x_0 = \tilde{\beta}\tan\alpha/\sqrt{n_2}$ and from Eq. (1.3-26), we have

$$x^2(z) + y^2(z) = x_0^2. \qquad (1.3\text{-}27)$$

The ray spirals around the z-axis as a helix. Figure 1.7 shows a MATLAB output for the m-file presented in Table 1.1. For $n(x_0,0) = 1.5, n_2 = 0.001,$ and a launching angle α of 0.5 radian, we have $x_0 = 22.74\mu$m.

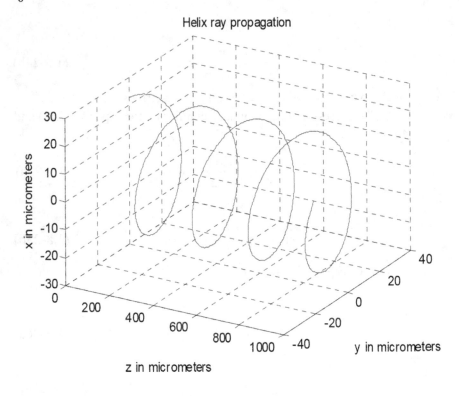

Fig. 1.7 Helix ray propagation.

Table 1.1 Helix.m: m-file for plotting helix ray propagation, and its corresponding output for the input parameters used.

```
%Helix.m
%Plotting Eq. (1.3-27)
clear
nxo = input('n(xo) = '); n2 = input('n2 = ');
alpha = input('alpha [radian] = ');
zin =input('start point of z in micrometers =  ');
zfi = input('end point of z in micrometers =  ');
Beta = nxo*cos(alpha);
z=zin:(zfi-zin)/1000:zfi;
xo=Beta*tan(alpha)/(n2^0.5);
x=xo*cos((n2^0.5)*z/Beta);
y=xo*sin((n2^0.5)*z/Beta);
plot3(z,y,x)
title('Helix ray propagation')
xlabel('z in micrometers')
ylabel('y in micrometers')
zlabel('x in micrometers')
grid on
sprintf('%f [micrometers]', xo)
view(-37.5+68, 30)
```

----

n(xo) = 1.5 n2 = 0.001 alpha [radian] = 0.5 start point of z in micrometers = 0 end point of z in micrometers = 1000
ans = 22.741150 [micrometers]

==

Ray Propagation through Ionosphere-like Medium Example

We consider a medium having a linear variation of the refractive index given below:

$$n^2(x) = n_0^2 - n_1 x \quad x > 0$$
$$= n_0^2 \qquad x \leq 0.$$

Let us find $x(z)$, in the region when $x > 0$, for a ray passing through $x = 0, z = 0$ with an angle α with respect to the optical axis, where $n_1 > 0$.

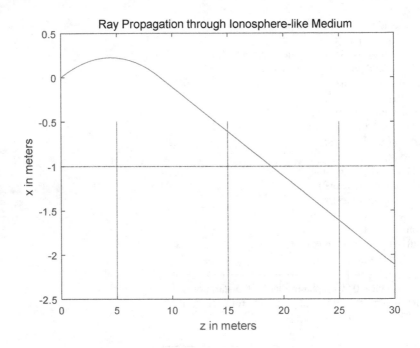

Ray propagation through a medium of a linear variation of the refractive index.

```
%Ray Propagation through Ionosphere-like Medium

clear
no = input('n(o) = ');
n1 = input('n1 = ');
alpha = input('alpha [radian] = ');
zfi = input('End of z in meters = ');
Beta=no*cos(alpha);
z=0:0.1:zfi;
x=-n1/(4*Beta^2)*z.^2+z*tan(alpha);
[abs_x z1_ind]=min(abs(x(2:length(x))));
z1_ind=z1_ind+1;
z_xneg=z(z1_ind:length(z));
```

```
x_neg= -z_xneg*tan(alpha)+(4*Beta^2*tan(alpha)^2)/n1;
x(z1_ind:length(z))=x_neg;
z0=z(z1_ind)
plot(z,x)
title('Ray Propagation through Ionosphere-like Medium ')
xlabel('z in meters')
ylabel('x in meters')
grid on
```

----

no= 1.5 n1 = 0.1 alpha [radian] = 0.1 and z1 (distance along z from origin when the ray crosses x=0) = 8.9 m

----

Ionosphere-like-medium.m: m-file for plotting ray propagation through ionosphere-like medium, and its corresponding output for the input parameters used.

From Eq. (1.3-20), we have

$$\frac{d^2x}{dz^2} = \frac{1}{2\tilde{\beta}^2} \frac{\partial n^2}{\partial x} = \frac{-n_1}{2\tilde{\beta}^2},$$

where $\tilde{\beta} = n(0)\cos\theta(0) = n_0 \cos\alpha$. The general solution to the above differential equation is

$$x(z) = \frac{-n_1}{4\tilde{\beta}^2} z^2 + Cz + D.$$

The corresponding initial conditions for the above equation are

$$x(0) = 0, \quad \frac{dx(0)}{dz} = \tan\alpha,$$

and the complete solution is

$$x(z) = \frac{-n_1}{4\tilde{\beta}^2} z^2 + z\tan\alpha.$$

Similarly, in the region when $x < 0$, we can find that the ray path is a straight line given by

$$x(z) = -z \tan \alpha + \frac{4 \tilde{\beta}^2 \tan^2 \alpha}{n_1}.$$

===

1.4 Matrix Methods in Paraxial Optics

Matrices may be used to describe ray propagation through optical systems comprising, for instance, a succession of spherical refracting and/or reflecting surfaces all centered on the same axis - the *optical axis*. We take the optical axis to be along the z-axis, which is also the general direction in which the rays travel. We will not consider skew rays and our discussion is only confined to those rays that lie in the x-z plane and that are close to the z-axis (called *paraxial rays*). Paraxial rays are close to the optical axis such that their angular deviation from it is small; hence, the sine and tangent of the angles may be approximated by the angles themselves. The reason for this paraxial approximation is that all paraxial rays starting from a given object point intersect at another point after passage through the optical system. We call this point the image point. Nonparaxial rays may not give rise to a single image point. This phenomenon, which is called *aberration*, is outside the scope of this book. Paraxial optical imaging is also sometimes called *Gaussian optics* as it was Karl Friedrich Gauss (1777-1855) who laid the foundations of the subject.

A ray at a certain point along the x-axis can be specified by its "coordinates," which contains the information of the position of the ray and its direction. Given this information, we want to find the coordinates of the ray at another location further down the optical axis, by means of successive operator acting on the initial ray coordinates, with each operator characteristic of the optical element through which the ray travels along the optical axis. We can represent these operators by matrices. The advantage of this matrix formalism is that any ray can be tracked during its propagation through the optical system by successive matrix multiplications, which can be easily done on a computer. This representation of geometrical optics is widely used in optical element designs.

In what follows, we will first develop the matrix formalism for paraxial ray propagation and examine some of the properties of ray transfer matrices. We then consider some illustrative examples.

1.4.1 *The ray transfer matrix*

Consider the propagation of a paraxial ray through an optical system as shown in Fig. 1.8. Our discussion is confined to those rays that lie in the *xz*-plane and are close to the *z*-axis (the optical axis). A ray at a given cross-section or plane may be specified by its height *x* from the optical axis and by its angle θ or slope which it makes with the *z*-axis. The convention for the angle is that θ is measured in radians and is anti-clockwise positive measured from the *z*-axis. The quantities (x, θ) represent the coordinates of the ray for a given *z*-constant plane. However, instead of specifying the angle the ray makes with the *z*-axis, it is customary to replace the corresponding angle θ by $v = n\theta$, where *n* is the refractive index at the *z*-constant plane.

In Fig. 1.8, the ray passes through the input plane with *input ray coordinates* $(x_1, v_1 = n_1\theta_1)$, then through the optical system, and finally through the output plane with *output ray coordinates* $(x_2, v_2 = n_2\theta_2)$. In the paraxial approximation, the corresponding output quantities are linearly dependent on the input quantities. We can, therefore, represent the transformation from the input to the output in matrix form as

$$\begin{pmatrix} x_2 \\ v_2 \end{pmatrix} = \begin{pmatrix} A & B \\ C & D \end{pmatrix} \begin{pmatrix} x_1 \\ v_1 \end{pmatrix}. \tag{1.4-1}$$

The above *ABCD* matrix is called the *ray transfer matrix* or the *system matrix, $\mathcal{S}$*. The system matrix can be made up of many matrices to account for the effects of a ray passing through various optical elements. We can consider these matrices as operators successively acting on the input ray coordinates. We state that the determinant of the ray transfer matrix equals unity, i.e., $AD - BC = 1$. This will become clear after we derive the translation, refraction and reflection matrices.

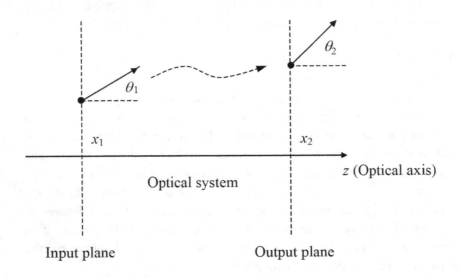

Fig. 1.8 Reference planes in an optical system.

Let us now attempt to understand better the significance of A, B, C and D by considering what happens if one of them vanishes within the ray transfer matrix.

a) If $D = 0$, we have from Eq. (1.4-1) that $v_2 = Cx_1$. This means that all rays crossing the input plane at the same point x_1, emerge at the output plane making the same angle with the axis, no matter at what angle they entered the system. The input plane is called the *front focal plane* of the optical system [see Fig. 1.9(a)].

b) If $B = 0, x_2 = Ax_1$ [from Eq.(1.4-1)]. This means that all rays passing through the input plane at the same point x_1 will pass through the same point x_2 in the output plane [see Fig. 1.9(b)]. The input and output planes are called the *object* and *image planes,* respectively. In addition, $A = x_2 / x_1$ gives the *magnification* produced by the system.

Furthermore, the two planes containing x_1 and x_2 are called *conjugate planes.* If $A = 1$, i.e., the magnification between the two conjugate planes is unity, these planes are called the *unit* or *principal* planes. The points of intersection of the unit planes with the optical axis are the *unit* or *principal points.* The principal points constitute one set of *cardinal points.*

c) If $C = 0, v_2 = Dv_1$. This means that all the rays entering the system parallel to one another will also emerge parallel, albeit in a new direction [see Fig. 1.9(c)]. In addition, $D(n_1 / n_2) = \theta_2 / \theta_1$ gives the *angular magnification* produced by the system.

If $D = n_2 / n_1$, we have unity angular magnification, i.e., $\theta_2 / \theta_1 = 1$. In this case, the input and output planes are referred to as the *nodal planes.* The intersections of the nodal planes with the optical axis are called the *nodal points* [see Fig. 1.9(d)]. The nodal points constitute a second set of cardinal points.

d) If $A = 0, x_2 = Bv_1$. This means that all rays entering the system at the same angle will pass through the same point at the output plane. The output plane is the *back focal plane* of the system [see Fig. 1.9(d)]. Note that the intersection of the front focal and back focal planes with the optical axis are called the *front and back focal points.* The focal points constitute the last set of cardinal points.

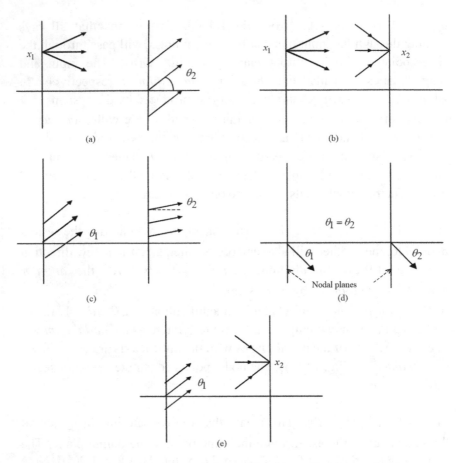

Fig. 1.9 Rays at input and output planes for (a) $D = 0$, (b) $B = 0$, (c) $C = 0$, (d) the case when the planes are nodal planes, and (e) $A = 0$.

Translation Matrix

Figure 1.10 shows a ray traveling a distance d in a homogeneous medium of refractive index n. Since the medium is homogeneous, the ray travels in a straight line [see Eq. (1.3-17)]. The set of equations of translation by a distance d is

$$x_2 = x_1 + d \tan \theta_1, \tag{1.4-2a}$$

and

$$n_2 \theta_2 = n \theta_1 \ or \ v_2 = v_1. \tag{1.4-2b}$$

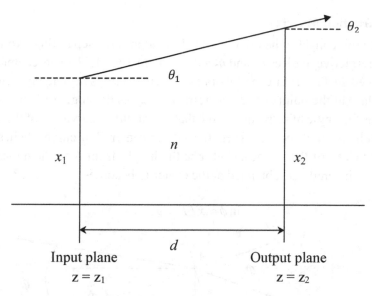

Fig. 1.10 A ray in a homogeneous medium of refractive index n.

From the above equations, we can relate the output coordinates of the ray with its input coordinates. We can express this transformation in a matrix form as

$$\begin{pmatrix} x_2 \\ v_2 \end{pmatrix} = \begin{pmatrix} 1 & d/n \\ 0 & 1 \end{pmatrix} \begin{pmatrix} x_1 \\ v_1 \end{pmatrix}. \qquad (1.4\text{-}3)$$

The 2×2 ray transfer matrix, for a translation distance of d in a homogeneous medium of refractive index n, is called the *translation matrix* $\mathcal{T}_d$:

$$\mathcal{T}_d = \begin{pmatrix} 1 & d/n \\ 0 & 1 \end{pmatrix}. \qquad (1.4\text{-}4)$$

Note that its determinant is unity.

Refraction Matrix

We now consider the effect of a spherical surface separating two regions of refractive indices n_1 and n_2 as shown in Fig. 1.11. The center of the curved surface is at C and its radius of curvature is R. The ray strikes the surface at the point A and gets refracted. ϕ_i is the angle of incidence and ϕ_t is the angle of refraction. Note that the radius of curvature of the surface will be taken as positive (negative) if the center C of curvature lies to the right (left) of the surface. Let x be the height from A to the optical axis. Then the angle ϕ subtended at the center C becomes

$$\sin \phi \approx x / R \approx \phi. \tag{1.4-5}$$

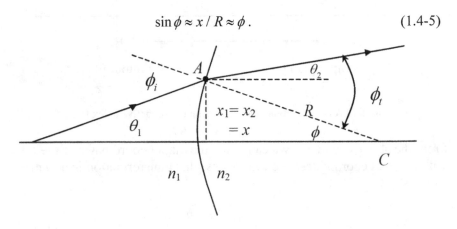

Fig. 1.11 Ray trajectory during refraction at a spherical surface.

We see that in this case, the height of the ray at A, before and after the refraction, is the same, i.e., $x_2 = x_1$. We therefore need to obtain the relationship for v_2 in terms of x_1 and v_1. Applying Snell's law [see Eq. (1.2-5)] and using the paraxial approximation, we have

$$n_1 \phi_i = n_2 \phi_t. \tag{1.4-6}$$

From geometry, we know from Fig. 1.11 that $\phi_i = \theta_1 + \phi$ and $\phi_t = \theta_2 + \phi$. Hence,

$$n_1 \phi_i = v_1 + n_1 x_1 / R, \tag{1.4-7a}$$

$$n_2 \phi_t = v_2 + n_2 x_2 / R. \tag{1.4-7b}$$

Using Eqs. (1.4-6), (1.4-7) and the fact that $x_1 = x_2$, we obtain

$$v_2 = \frac{n_1 - n_2}{R} x_1 + v_1. \qquad (1.4-8)$$

The matrix relating the coordinates of the ray after refraction to those before refraction becomes

$$\begin{pmatrix} x_2 \\ v_2 \end{pmatrix} = \begin{pmatrix} 1 & 0 \\ -p & 1 \end{pmatrix} \begin{pmatrix} x_1 \\ v_1 \end{pmatrix}, \qquad (1.4-9a)$$

where the quantity p given by

$$p = \frac{n_2 - n_1}{R} \qquad (1.4-9b)$$

is called the *refracting power* of the spherical surface. When R is measured in meters, the unit of p is called *diopters*. If an incident ray is made to converge (diverge) by a surface, the power will be assumed to be positive (negative) in sign. The (2×2) transfer matrix is called the *refraction matrix* $\mathfrak{R}$ and it describes refraction for the spherical surface:

$$\mathfrak{R} = \begin{pmatrix} 1 & 0 \\ \dfrac{n_1 - n_2}{R} & 1 \end{pmatrix}. \qquad (1.4-10)$$

Note that the determinant of $\mathfrak{R}$ is also unity.

Thin-Lens Matrix

Consider a thick lens as shown in Fig. 1.12. We can show that the input ray coordinates (x_1, v_1) and the output ray coordinates (x_2, v_2) are connected by three matrices (a refraction matrix followed by a translation matrix and then by another refraction matrix):

$$\begin{pmatrix} x_2 \\ v_2 \end{pmatrix} = \mathcal{S} \begin{pmatrix} x_1 \\ v_1 \end{pmatrix}, \qquad\qquad (1.4\text{-}11)$$

where $\mathcal{S}$ is the system matrix and given by, using Eqs. (1.4-4) with $n = n_2$ and Eq. (1.4-10)

$$\mathcal{S} = \mathfrak{R}_2 \mathcal{T}_d \mathfrak{R}_1$$

$$= \begin{pmatrix} 1 & 0 \\ \dfrac{n_2 - n_1}{R_2} & 1 \end{pmatrix} \begin{pmatrix} 1 & d/n_2 \\ 0 & 1 \end{pmatrix} \begin{pmatrix} 1 & 0 \\ \dfrac{n_1 - n_2}{R_1} & 1 \end{pmatrix}.$$

refraction at   translation refraction at
surface 2   surface 1

Note that in $\mathfrak{R}_2$, we have interchanged n_1 and n_2 to take into the account that the ray is traveling from n_2 to n_1.

     We see that in general, the system matrix is a product of any number of 2×2 matrices, and the system matrix is itself a 2×2 matrix. Now, from linear algebra, we know that the determinant of a product matrix, such as $\mathcal{S}$, is the product of the individual determinants. Hence, the determinant of a system matrix is unity. It is, therefore, useful in checking the correctness of the calculations that produce a system matrix, which has a unit determinant. However, the condition of a unit determinant is a necessary but not a sufficient condition on the system matrix as an arbitrary 2×2 matrix with unit determinant does not necessarily correspond to a real physical system.

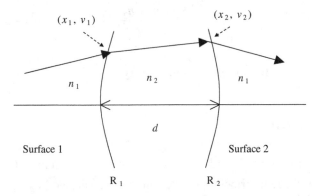

Fig. 1.12 A thick lens: The radii of curvatures of surfaces 1 and 2 are R_1 and R_2.

For an ideal thin lens in air, $d \to 0$ and $n_1 = 1$. Writing $n_2 = n$ for notational convenience, Eq.(1.4-11) becomes

$$S = \begin{pmatrix} 1 & 0 \\ -p_2 & 1 \end{pmatrix} \begin{pmatrix} 1 & 0 \\ 0 & 1 \end{pmatrix} \begin{pmatrix} 1 & 0 \\ -p_1 & 1 \end{pmatrix}, \qquad (1.4\text{-}12)$$

where $p_1 = (n-1)/R_1$ and $p_2 = (1-n)/R_2$ are the *refracting powers* of surfaces 1 and 2, respectively. Note that the translation matrix degenerates into a unit matrix. Equation (1.4-12) can be rewritten as

$$S = \begin{pmatrix} 1 & 0 \\ -p_2 & 1 \end{pmatrix} \begin{pmatrix} 1 & 0 \\ -p_1 & 1 \end{pmatrix} = \begin{pmatrix} 1 & 0 \\ -1/f & 1 \end{pmatrix} = \mathcal{L}_f, \qquad (1.4\text{-}13)$$

where $\mathcal{L}_f$ is called the *thin-lens matrix* and f is the *focal length* of the lens given by

$$\frac{1}{f} = (n-1)\left(\frac{1}{R_1} - \frac{1}{R_2} \right). \qquad (1.4\text{-}14)$$

For $R_1 > (<) \, 0$ and $R_2 < (>) \, 0$, we have $f > (<) \, 0$. If a ray of light is incident on the left surface of the lens parallel to the axis and for $f > (<) \, 0$, the ray bends towards (away from) the axis upon refraction through the lens. In the first case, the lens is called a *converging* (*convex*) lens, while in the second case, we have a *diverging* (*concave*) lens.

1.4.2 Illustrative examples

Example 1.4 Ray Tracing through a Single Thin Lens

(a) Ray traveling parallel to the optical axis:
The input ray coordinates are $(x_1, 0)$, and hence the output ray coordinates are given, using Eqs. (1.4-1) and (1.4-13), as $(x_1, -x_1/f)$. This ray now travels in a straight line at an angle $-1/f$ with the axis, which means that

if x_1 is positive (or negative), the ray after refraction through the lens intersects the optical axis at a point a distance f behind the lens if the lens is converging ($f > 0$). This justifies why f is called the focal length of the lens. All rays parallel to the optical axis in front of the lens converge behind the lens to a point called the *back focus* [see Fig. 1.13(a)]. In the case of a diverging lens ($f < 0$), the ray after refraction diverges away from the axis as if it were coming from a point on the axis a distance f in front of the lens. This point is called the *front focus*. This is also shown in Fig. 1.13(a).

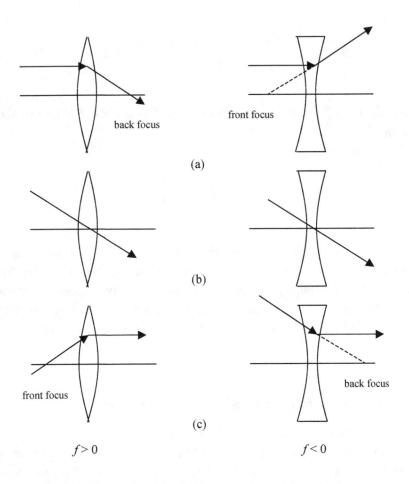

Fig. 1.13 Ray tracing through thin converging and diverging lenses.

(b) Ray traveling through the center of the lens:

The input ray coordinates are $(0, v_1)$, and hence the output ray coordinates are given, using Eqs. (1.4-6) and (1.4-13), as $(0, v_1)$, which means that a ray traveling through the center of the lens will pass undeviated as shown in Fig. 1.13(b).

(c) Ray passing through the front focus of a converging lens:

The input ray coordinates are given by $(x_1, x_1 / f)$, so that the output ray coordinates are $(x_1, 0)$. This means that the output ray will be parallel to the axis, as shown in Fig. 1.13(c).

In a similar way, we can also show that for an input ray on a diverging lens appearing to travel toward its *back focus*, the output ray will be parallel to the axis.

Example 1.5 Imaging by a Single Thin Lens

Consider an object OO' located a distance in front of a thin lens of focal length f, as shown in Fig. 1.14.

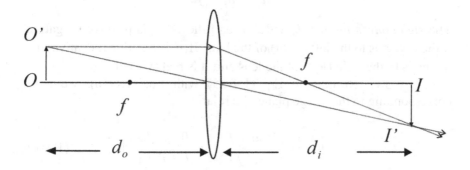

Fig. 1.14 Imaging by a single lens.

Assume that (x_0, v_0) represents the input ray coordinates originally from point O', and traveling towards the lens for a distance of d_0. Then the output ray coordinates (x_i, v_i) at a distance d_i behind the lens can be written in terms of the input ray coordinates, two translation matrices for air ($n=1$) [see Eq.(1.4-4)] and the thin-lens matrix [see Eq. (1.4-13)] as:

$$\begin{pmatrix} x_i \\ v_i \end{pmatrix} = T_{d_i} \mathcal{L}_f T_{d_0} \begin{pmatrix} x_0 \\ v_0 \end{pmatrix}$$

$$= \begin{pmatrix} 1 & d_i \\ 0 & 1 \end{pmatrix} \begin{pmatrix} 1 & 0 \\ -1/f & 1 \end{pmatrix} \begin{pmatrix} 1 & d_0 \\ 0 & 1 \end{pmatrix} \begin{pmatrix} x_0 \\ v_0 \end{pmatrix}$$

$$= \begin{pmatrix} 1-d_i/f & d_0+d_i-d_0 d_i/f \\ -1/f & 1-d_0/f \end{pmatrix} \begin{pmatrix} x_0 \\ v_0 \end{pmatrix}$$

$$= \mathcal{S} \begin{pmatrix} x_0 \\ v_0 \end{pmatrix}.$$

(1.4-15)

We see that $\mathcal{S}$ is the system matrix in our case and by setting $B=0$ [see Eq. (1.4-10)] in the matrix we have the following celebrated *thin-lens formula* for the imaging lens:

$$\frac{1}{d_0} + \frac{1}{d_i} = \frac{1}{f}.$$

(1.4-16)

The *sign convention* for d_0 and d_i is as follows. d_0 is positive (negative) if the object is to the left (right) of the lens. If d_i is positive (negative), the image is to the right (left) of the lens and it is real (virtual).

Now, returning to Eq. (1.4-15) with Eq. (1.4-16), we have, corresponding to the image plane, the relation

$$\begin{pmatrix} x_i \\ v_i \end{pmatrix} = \begin{pmatrix} 1-d_i/f & 0 \\ -1/f & 1-d_0/f \end{pmatrix} \begin{pmatrix} x_0 \\ v_0 \end{pmatrix}.$$

(1.4-17)

For $x_0 \neq 0$, we obtain

$$\frac{x_i}{x_0} = M = 1 - \frac{d_i}{f} = \frac{f-d_i}{f} = \frac{f}{f-d_0} = -\frac{d_i}{d_0}$$

(1.4-18)

using Eq. (1.4-16), where M is called the *lateral magnification* of the system. If $M > 0$ (< 0), the image is erect (inverted).

==

Volumetric Imaging of a Single Lens Example

With reference to the figure in the Example, we investigate imaging of a volume with a single lens.

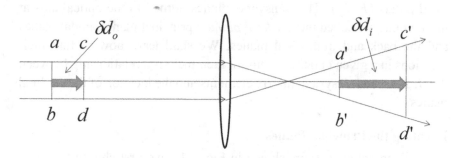

Volume imaging with a single lens.

We need to consider longitudinal magnification in addition to lateral magnification. Longitudinal magnification M_Z is the ratio of an image displacement along the axial direction, δd_i, to the corresponding object displacement, δd_0, i.e., $M_Z = \delta d_i / \delta d_0$. Using Eq. (1.4-16) ad treating d_i and d_0 as variables, we take the derivative of d_i with respect to d_0 to obtain

$$M_Z = \delta d_i / \delta d_0 = -M^2.$$

The above equation states that the longitudinal magnification is equal to the square of the lateral magnification. The minus sign in front of the equation signifies that a decrease in the distance of the object from the lens, $|d_0|$, will result in an increase in the image distance, $|d_i|$. The situation of a magnified volume is shown in the figure above, where a cube

volume (abcd plus the dimension into the paper) is imaged into a truncated
pyramid with a-b imaged into a'-b' and c-d imaged into c'-d'.
==

1.4.3 *Cardinal points of an optical system*

We have briefly mentioned cardinal points in Section 1.4.1 and pointed
out that there are six *cardinal points* on the optical axis that characterize
an optical system. They are the first and second principal (unit) points
(H_1, H_2), first and second nodal points (N_1, N_2) and the front and back
focal points (F_1, F_2). The transverse planes normal to the optical axis at
these points are called the *cardinal planes:* principal planes, nodal planes
and the back and front focal planes. We shall learn how to find their
locations in a given optical system. In fact, there is a relationship between
the A, B, C, and D system matrix elements and the location of the cardinal
planes.

Locating the Principal Planes

For a given optical system, shown in Fig. 1.15, we first choose the input
plane and the output plane. We then assume that we know the *ABCD*
system matrix linking the two chosen planes. Now for the sake of
generality, we take n_1 and n_2 to be the refractive indices to the left and to
the right of the two planes, respectively.

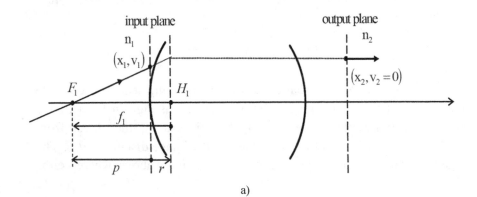

a)

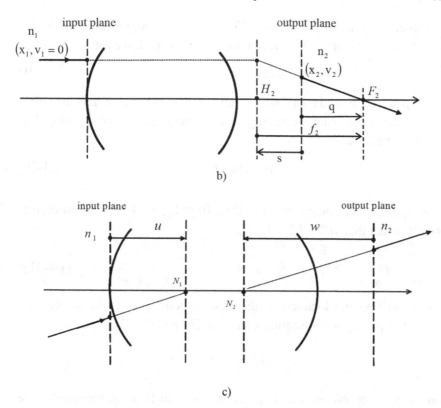

Fig. 1.15. (a) Ray crossing first focal point F_1 is bent parallel to the optical axis at the first principal plane, (b) Ray entering the system parallel to the optical axis is bent at the second principal plane in such a way that it passes through the second focal point F_2, and (c) Ray entering the system directed towards N_1 is emerged as a ray coming from N_2.

Consider first, as in Fig. 1.15(a), a ray crossing F_1, by definition, is bent parallel to the optical axis at the first principal plane. The focal point is located at a distance f_1 from the principal plane and at a distance p from the input plane. Furthermore, the distance r locates the principal point from the input plane. The convention for distances are that distances measured to the right of their planes are considered positive and to the left, negative. Since the input ray coordinates and the output ray coordinates are related by the *ABCD* matrix given by Eq. (1.4-1), we can write

$$v_2 = Cx_1 + Dn_1\theta_1 = 0, \ (v_1 = n_1\theta_1). \tag{1.4-19}$$

Now, $p = -x_1 / \theta_1$ and the negative sign is included because p is to the left side of the input plane according to the convention. Incorporating Eq. (1.4-19), we have

$$p = Dn_1 / C. \tag{1.4-20}$$

Now $f_1 = -x_2 / \theta_1$ where $x_2 = Ax_1 + Bn_1\theta_1$ from Eq. (1.4-1). We can derive, using the fact that $AD - BC = 1$,

$$f_1 = n_1 / C. \tag{1.4-21}$$

Finally to find the location of the first principal point, we notice that $r = -(f_1 - p)$. By incorporating Eqs. (1.4-20) and (1.4-21), we have

$$r = n_1(D-1)/C. \tag{1.4-22}$$

Similarly, with reference to Fig. 1.15(b) we can find the location of the second principal plane. By definition, a ray which enters the system parallel to the optical axis at the height x_1 arrives the same height at the second principal plane. The ray is then bent at the second principal plane in such a way that it passes through the second focal point F_2. Again, the convention for distances q, s, and f_2 are that distances measured to the right of their planes (output plane and the second principal plane) are considered positive and to the left, negative. We, therefore, write that

$$q = -x_2 / \theta_2, \quad (v_2 = n_2\theta_2), \tag{1.4-23}$$

where the negative sign is included in the above equation as $\theta_2 < 0$. Now, from Eq. (1.4-1), we have $v_2 = Cx_1$ and $x_2 = Ax_1$, and we can rewrite Eq. (1.4-23) in terms of the elements of the *ABCD* matrix:

$$q = -An_2 / C. \tag{1.4-24}$$

To find the second focal length, we write $f_2 = -x_1 / \theta_2$. Using $v_2 = Cx_1$, the second focal length is

$$f_2 = -n_2 / C. \tag{1.4-25}$$

Finally, to find s, we refer to Fig. 1.15(b) and write $s = q - f_2$. Using Eqs. (1.4-24) and (1.4-25), we have

$$s = n_2(1 - A)/C. \tag{1.4-26}$$

Locating the Nodal Planes

Similarly, we can find the location of the Nodal planes with reference to Fig. 1.15(c). Again, the convention for distances u and w are that distances measured to the right of their planes (output plane and input planes) are considered positive and to the left, negative. We state the results as follows:

$$u = (Dn_1 - n_2)/C, \tag{1.4-27}$$

and

$$w = (n_1 - An_2)/C. \tag{1.4-28}$$

Example 1.6 Ray Tracing Using Principal Planes and
Nodal Planes

An object (an erected arrow denoted by O) is located $20cm$ from the ideal positive lens of focal length $f_p = 10cm$. The distance between the positive lens and the ideal negative lens of focal length $f_p = -10cm$ is 5 cm as shown in Fig. 1.16. We shall draw a ray diagram for the two lens optical system for the image (I).

We choose the input and output planes to be the location where the positive lens and the negative lens are situated, respectively. The system matrix $\mathcal{S}$ linking the two planes is

$$\mathcal{S} = \begin{pmatrix} 1 & 0 \\ 10 & 1 \end{pmatrix}\begin{pmatrix} 1 & 0.05 \\ 0 & 1 \end{pmatrix}\begin{pmatrix} 1 & 0 \\ -10 & 1 \end{pmatrix} = \begin{pmatrix} 0.5 & 0.05 \\ -5 & 1.5 \end{pmatrix}, \tag{1.4-29}$$

where we have converted distances to meters and focal lengths to diopters.

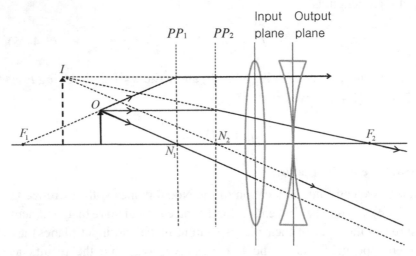

Fig. 1.16 Ray tracing using principal planes and nodal planes.

Since we have found the $ABCD$ matrix of the system, we can now find all the cardinal points and planes, which will help us to draw a ray diagram for the optical system. Assuming $n_1 = n_2 = 1$, i.e., the optical system is immersed in air, we tabulate the results as follows:

Front focal point F_1:
$$p = D/C = 1.5/(-5) = -0.3\,m$$
$$= -30\,cm < 0 \; (30\,cm \text{ left of the input plane}).$$

Back focal point F_2:
$$q = -A/C = -0.5/(-5) = 0.1\,m$$
$$= 10\,cm > 0 \; (10\,cm \text{ right of the output plane}).$$

First Principal point H_1:
$$r = (D-1)/C = (1.5-1)/(-5) = -0.1\,m$$
$$= -10\,cm < 0$$
($10\,cm$ left the input plane, PP_1 denotes the first principal plane).

Second Principal point H_2 :

$$s = (1 - A) / C = (1 - 0.5) / (-5) = -0.1 \, m$$
$$= -10 \, cm < 0$$

(10 *cm* left of the output plane, PP_2 denotes the second principal plane).

First Nodal point N_1 :

$$u = (D - 1) / C$$
$$= -10 \, cm < 0 \, (10 \, cm \text{ left of the input plane}).$$

Second Nodal point N_2 :

$$w = (1 - A) / C$$
$$= -10 \, cm < 0 \, (10 \, cm \text{ left of the output plane}).$$

Notice that the equivalent focal length of the optical system is

$$f_2 = -1 / C = 20 \, cm.$$

1.5 Reflection Matrix and Optical Resonators

There is a rule that will enable us to use the translation matrix $\mathcal{T}_d$ and refraction matrix $\mathfrak{R}$ even for reflecting surfaces such as mirrors. When a light ray is traveling in the $-z$ direction, the refractive index of the medium through which the ray is transversing is taken as negative. According to the rule, from the refraction matrix [see Eq. (1.4-10)], we can modify it to become the *reflection matrix* $\tilde{\mathfrak{R}}$:

$$\tilde{\mathfrak{R}} = \begin{pmatrix} 1 & 0 \\ -p & 1 \end{pmatrix},$$

where $p = \dfrac{n_2 - n_1}{R} = \dfrac{(-n_1) - n_1}{R} = \dfrac{-2n}{R}$ and n is the refractive index for the medium in which the mirror is immersed. The situation is shown in Fig. 1.17. Hence we can write the reflection matrix as

$$\tilde{\mathfrak{R}} = \begin{pmatrix} 1 & 0 \\ \dfrac{2n}{R} & 1 \end{pmatrix}. \qquad (1.5\text{-}1)$$

We see that if the rule is used on the equation for the power of a surface, we find that a concave mirror (R being negative) will give a positive power p, which is in agreement with the common knowledge that a concave mirror will focus rays, as illustrated in Fig. 1.17. The focal length of the spherical mirror is $f = -R / 2n$.

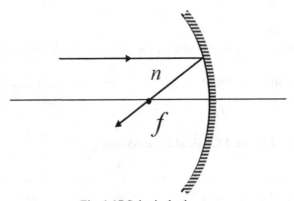

Fig. 1.17 Spherical mirror.

In dealing with the translation matrix upon ray reflection, we adopt the convention in that when light rays travel between planes $z = z_1$ and $z = z_2 > z_1, z_1 - z_2$ is taken to be positive (negative) for a ray traveling in the $+z(-z)$ direction. Again, the refractive index of the medium is taken to be negative. By taking the value of the refractive index to be negative when a ray is traveling in the $-z$ direction, we can use the same translation matrix throughout the analysis when reflecting surfaces are included in the optical system. With reference to Fig. 1.18, the translation matrices between various planes are given as follows:

$$T_{12} = \begin{pmatrix} 1 & d/n \\ 0 & 1 \end{pmatrix} \qquad \text{between planes 1 and 2;}$$

$$T_{32} = \begin{pmatrix} 1 & -d/-n \\ 0 & 1 \end{pmatrix} \qquad \text{between planes 2 and 3,}$$

and

$$T_{32} = T_{32}T_{21} = \begin{pmatrix} 1 & 2d/n \\ 0 & 1 \end{pmatrix} \qquad \text{between planes 1 and 3.}$$

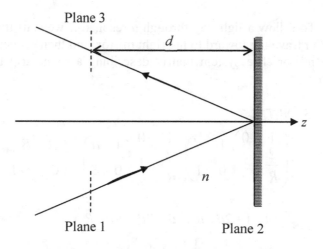

Fig. 1.18 Rays reflected from a plane mirror.

Optical Resonators

An *optical resonator* is an optical system consisting of two mirrors of radii of curvature R_1 and R_2, separated by a distance d, as shown in Fig. 1.19. The resonator forms an important part of a laser system. Indeed, for sustained oscillations, implying a constant laser output, the resonator must be stable. We shall now obtain the condition for a stable resonator. In stable resonators, a light ray must keep bouncing back and forth and remain trapped inside in order that oscillations are sustained.

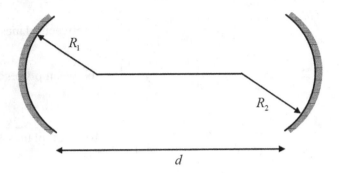

Fig. 1.19 Resonator consisting of two spherical mirrors.

To follow a light ray through a resonator, we start the ray at the left mirror traveling toward to the right mirror, and then reflecting back to the left mirror. The system matrix describing a round trip through the resonator is

$$\mathcal{S} = \tilde{\mathfrak{R}}_1 T_d \tilde{\mathfrak{R}}_2 T_d$$

$$= \begin{pmatrix} 1 & 0 \\ \dfrac{2}{R_1} & 1 \end{pmatrix} \begin{pmatrix} 1 & d \\ 0 & 1 \end{pmatrix} \begin{pmatrix} 1 & 0 \\ \dfrac{2}{R_2} & 1 \end{pmatrix} \begin{pmatrix} 1 & d \\ 0 & 1 \end{pmatrix} = \begin{pmatrix} A & B \\ C & D \end{pmatrix}, \qquad (1.5\text{-}2a)$$

where

$$A = 1 + 2d/R_2, \quad B = 2d(1 + d/R_2),$$

$$C = 2\left[\frac{1}{R_1} + \frac{1}{R_2}\left(1 + \frac{2d}{R_1}\right)\right], \qquad (1.5\text{-}2b)$$

$$D = \frac{2d}{R_1} + \left(1 + \frac{2d}{R_1}\right)\left(1 + \frac{2d}{R_2}\right).$$

Hence, we can write

$$\begin{pmatrix} x_1 \\ y_1 \end{pmatrix} = \begin{pmatrix} A & B \\ C & D \end{pmatrix} \begin{pmatrix} x_0 \\ v_0 \end{pmatrix},$$

where (x_1, v_1) is the ray coordinates after one round trip and (x_0, v_0) is the ray coordinates when it started from the left mirror. Now, the coordinates of the ray (x_m, v_m) after m complete round trips (oscillations) would be

$$\begin{pmatrix} x_m \\ v_m \end{pmatrix} = \begin{pmatrix} A & B \\ C & D \end{pmatrix}^m \begin{pmatrix} x_0 \\ v_0 \end{pmatrix}. \tag{1.5-3}$$

We can show that

$$\begin{pmatrix} A & B \\ C & D \end{pmatrix}^m = \frac{1}{\sin\theta} \begin{pmatrix} A\sin m\theta - \sin(m-1)\theta & B\sin m\theta \\ C\sin m\theta & D\sin(m-1)\theta \end{pmatrix}, \tag{1.5-4}$$

where the angle θ has been defined as

$$\cos\theta = \frac{1}{2}(A+D). \tag{1.5-5}$$

In order to achieve stability, the coordinates of the ray after m trips should not diverge as $m \to \infty$. This happens if the magnitude of $\cos\theta$ is less than 1. In other words, if θ is a complex number, the terms $\sin m\theta$ and $\sin(m-1)\theta$ in Eq. (1.5.4) diverges. Hence the *stability criterion* is

$$-1 \le \cos\theta \le 1 \tag{1.5-6}$$

or, when using Eqs. (1.5-5) and (1.5-2b),

$$0 \le \left(1 + \frac{d}{R_1}\right)\left(1 + \frac{d}{R_2}\right) \le 1. \tag{1.5-7}$$

The stability criterion is often written using the so-called g *parameters* of the resonator as

$$0 \le g_1 g_2 \le 1, \tag{1.5-8}$$

where $g_1 = \left(1 + \dfrac{d}{R_1}\right)$ and $g_2 = \left(1 + \dfrac{d}{R_2}\right)$.

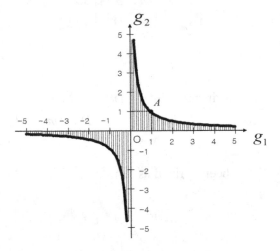

Fig. 1.20 Stability diagram for optical resonators.

Figure 1.20 shows the *stability diagram* for optical resonators. Only those resonator configurations that lie in the shaded region correspond to a stable configuration. The point marked O corresponds to the so-called *confocal configuration*, where $R_1 = d, R_2 = -d$ or $g_1 g_2 = 0$. Figure 1.21 shows ray propagation inside such a resonator.

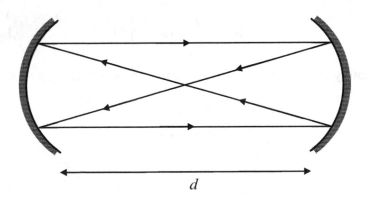

Fig. 1.21 Ray propagation inside a stable resonator.

1.6 Ray Optics using MATLAB

Example 1 Obtaining output ray coordinates of a single lens system

We shall find the ray coordinates $r_i = (x_i, v_i)$ an arbitrary distance z behind a lens of focal length f when the input ray coordinates $r_0 = (x_0, v_0)$ for a ray starting from an object located a distance d_0 in front of the lens is specified. $\mathcal{T}_{d_O}$ and $\mathcal{T}_z$ denote the translation matrices for the ray in air before and after the lens (corresponding to object and image distances, respectively), while $\mathcal{L}_f$ is the lens matrix. The product of the three $S = \mathcal{T}_z \mathcal{L}_f \mathcal{T}_{d_O}$ gives the overall system matrix for the optical system. The program gives the output ray coordinates r_i. All distance dimensions have been written in centimeters.

As an example, we create the MATLAB function Ray_s. In MATLAB, after the prompt >>, we type [detS,ri] = Ray_s([0;1],15,10,30) to denote input conditions: $r_O = (0, 1)$, $d_O = 15\,cm$, $f = 10\,cm$, and $z = 30\,cm$. We obtain

$$r_i = (0, -0.5)$$

as an output.

Table 1.2 MATLAB code for ray traveling through a single lens, and the corresponding MATLAB output.

```
function [detS, ri]=Ray_s(ro, do, f,z); %This function is for output ray vector of
%a single lens system
To=[1, do;0,1];
Lf=[1,0;-(1/f),1];
Ti=[1,z;0,1];
S=Ti*Lf*To;
%Checking determinant for overall matrix
detS=det(S);
%"image" ray coordinate is ri
ri=S*ro;
```

```
Type in Matlab prompt
>>[detS, ri]=Ray_s([0;1], 15, 10, 30)
```

Output from Matlab

detS =   1

ri = 0

   -0.5000

We interpret the program as follows: if the input ray starts from the optical axis at a distance of 15 cm from the lens with $v = 1$ rad, the output ray meets the axis a distance of 30 cm behind the single lens with $v = -0.5$ rad. In other words, for an object distance d_O of 15 cm, the image distance $z = d_i$ is 30 cm. Finding the determinant of the overall system matrix S being unity is a check of the computations. Note that the ray coordinates at any plane z behind the lens can be found by substituting a number for the value of z in the program.

　　　To find the lateral magnification of the imaging system, we can enter in input ray coordinates of, say, (1,1). Using the same program as above, the output ray coordinates at $z = 30$ cm (the image plane) works out to be (−2.0, −0.6). This means that the magnification of the system equals −2, which corresponds to an inverted real image of twice the size as the object, as expected.

Example 2 *Obtaining output ray coordinates of a single lens system*

We shall find the ray coordinates $r_i = (x_i, v_i)$ an arbitrary distance z behind a two-lens combination of focal lengths $f_{1,2}$ and separated by a distance d . When the input ray coordinates r_o for rays starting from an object located a distance d_O in front of the lens are specified. T_{d_O} and T_{d_i} denote the translation matrices for the ray in air before and after the lens (corresponding to object and image distances, respectively). T_d denotes the translation matrix for a ray traveling between the two lenses, while

$\mathcal{L}_{f_{1,2}}$ are the lens matrices, respectively. The product of $S = T_{d_i} \mathcal{L}_{f_2} T_d \mathcal{L}_{f_1} T_{d_o}$ gives the overall system matrix for the optical system. The program gives the output ray coordinate r_i. All dimensions have been written in centimeters.

We create the MATLAB function Ray_d, as shown in Table 1.3. In MATLAB, after the prompt >>, we type [detS,ri]= Ray_d([0;1],10,10,10,10,20) to denote input conditions of the function:

$r_o = (0,1), d_o = 10 \, cm, z = 10 \, cm, f_1 = 10 \, cm, f_2 = 10 \, cm$ and $d = 20 \, cm$. We obtain output coordinates $r_i = (0, -1)$ as an output. Note also that when we input $r_o = (1, 0)$ with other input variables the same, we get $r_i = (-1, 0)$. This corresponds that when the input ray is parallel to the optical axis, the output ray is also parallel to the axis with an inverted image having a unit magnification.

Table 1.3   MATLAB code for ray traveling through a two lens system, and the corresponding MATLAB output.

```
function [detS, ri]=Ray_d(ro, do, z, f1, f2, d);
%This function is for output ray vector of a double lens system
To=[1, do;0,1];
Lf1=[1,0;-(1/f1),1];
Td=[1,d;0,1];
Lf2=[1,0;-(1/f2),1];
Ti=[1,z;0,1];
S=Ti*Lf2*Td*Lf1*To;

%Checking determinant for overall matrix
detS=det(S);
%"image" ray coordinate is ri
ri=S*ro;
```
detS =
   1
ri =
0
   -1

Example 3 Finding the image location in a single lens system

The following program, as shown in Table 1.4, is an extension to the first MATLAB example in this section. The MATLAB function Ray_z gives the location of the image plane for a given location of the object in a single imaging lens system. To do this, the object is taken to be an on-axis point, and the ray coordinates monitored behind the lens. If the position of the output ray is sufficiently close to the optical axis behind the lens, the corresponding value of z is the location of the image. As an example, we input $d_o = 15\,cm$, $f = 10\,cm$, $Z_s = 0$, $Z_f = 50\,cm$, and $\Delta z = 0.1\,cm$, where Z_s, Z_f and Δz represent the start and end points of the search range and the resolution of the search, respectively. The program output shows that indeed the image location works out to be a distance of 30 cm behind the lens for an object distance of 15 cm in front of the lens with focal length equal to 10 cm.

Table 1.4 MATLAB code for locating image plane for single lens imaging, and the corresponding MATLAB output.

```
function [z_est, M]=Ray_z(do, f, Zs, Zf, dz);
%This function is for searching image distance of the single lens system
To=[1, do;0,1]; Lf=[1,0;-
(1/f),1];

ro=[0;1]; n=0;
for z=Zs:dz:Zf
n=n+1;
Z1(n)=z;
  Ti=[1,z;0,1];
          S=Ti*Lf*To;
  %"image" ray coordinate is ri
          ri=S*ro;
  Ri(n)=ri(1,1);
end
[M, N]=min(abs(Ri));
z_est=Z1(N);
```

```
>>[z_est, M]=Ray_z(15, 10, 0, 50, 0.1)
z_est =
   30
M =
   0
```

Problems

1.1   A laser rocket is accelerated in free space by a photon engine
         that emits 10 kW of blue light ($\lambda = 450\ nm$).
         a) What is the force on the rocket?
         b) If the rocket weighs 100 kg, what is its acceleration?
         c) How far will it have traveled in one year if it starts from
         zero velocity?
         [Courtesy of Adrian Korpel, Professor Emeritus, Univ. Iowa]

1.2   Derive the laws of reflection and refraction by considering the
         incident, reflected and refracted light to comprise a stream of
         photons characterized by a momentum $p = \hbar k$ and k is the
         wavevector in the direction of ray propagation. Employ the law
         of conservation of momentum, assuming that the interface,
         say, $y =$ constant, only affects the y-component of the
         momentum. This provides an alternative derivation of the laws
         of reflection and refraction.

1.3   Show that the z-component of the ray equation,

$$\frac{d}{ds}\left(n\frac{dz}{ds}\right) = \frac{\partial n}{\partial z} \ ,$$

         can be derived directly from the equations for x and y [Eq.
         (1.3-13)]. Hint: make use of

$$ds^2 = dx^2 + dy^2 + dz^2.$$

1.4   a) Show that for the square-law medium

$$n^2(x, y) = n_o^2 - n_2(x^2 + y^2),$$

         we have

$$x(z) = \frac{n_o \sin\alpha}{\sqrt{n_2}}\sin\left(\frac{\sqrt{n_2}}{n_o \cos\alpha}z\right)$$

when the initial ray position is at $x = 0$ with α being the launching angle.

(b) Plot $x(z)$ for $\alpha = 10°, 20°,$ and $30°$ when $n_o = 1.5$, $n_2 = 0.1 mm^{-2}$ Draw some conclusion from the plots.

(c) For paraxial rays, i.e., α is small, can you draw a different conclusion from that obtained from part (b)?

1.5   Show that the ray transfer matrix for the square-law medium $n^2(x, y) = n_o^2 - n_2(x^2 + y^2)$ is

$$
\begin{pmatrix}
\cos \beta z & \dfrac{1}{n_o \beta} \sin \beta z \\
-n_o \beta \sin \beta z & \cos \beta z
\end{pmatrix},
$$

where $\beta = \sqrt{n_2} / n_o$.

1.6   An object is placed a distance 2m away from a concave spherical mirror of radius of 80 cm. Find the location of the image and draw the ray diagram of the image formed. Also find the magnification of the imaging system.

1.7   With reference to the volume imaging of a single lens example [see p. 37], show that

$$
M_z = -M^2,
$$

where M_z is the longitudinal magnification and M is the lateral magnification.

1.8   For medium

$$n^2(x) = n_0^2 - n_1 x \quad x > 0$$
$$= n_0^2 \qquad x \le 0,$$

we have a parabolic trajectory of a ray path shown below in Fig. P1.8, where α is the launching angle at $z = 0$. These ray paths are good examples for radio wave propagation through the ionosphere.

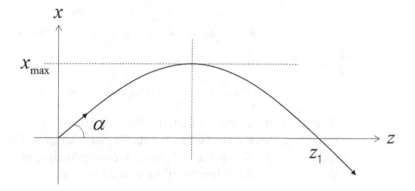

Fig. P1.8 Ray propagation through a medium of a linear variation of the refractive index.

Show that

$$x_{max} = \frac{\tilde{\beta}^2 \tan^2 \alpha}{n_1} \quad \text{and} \quad z_1 = \frac{4\tilde{\beta}^2 \tan \alpha}{n_1},$$

where z_1 is the distance from $z = 0$ to the point where the ray crosses $x = 0$ and $\tilde{\beta} = n_0 \cos \alpha$.

1.9 An object is placed 12 cm in front of a lens-mirror combination as shown in Fig. P1.9. Using ray transfer matrix concepts, find the position, and magnification of the image. Also, draw the ray diagram of the optical system.

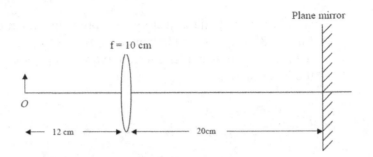

Fig. P1.9 Lens — Mirror combination

1.10 A glass hemisphere, shown in Fig. P1.10, of radius r and refractive index n is used as a lens for paraxial rays. Find the location of the first principal plane, second principal plane and the equivalent focal length of the optical system.

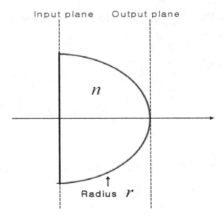

Fig. P1.10 Glass hemisphere as a thick lens

1.11 Referring to Fig. P1.11, show that the equivalent focal length f of the two-lens combination can be expressed as

$$\frac{1}{f} = \frac{1}{f_a} + \frac{1}{f_b} - \frac{d}{f_a f_b},$$

assuming $d < f_a + f_b$. Also, locate the second principal plane from which where f is measured.

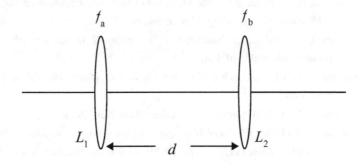

Fig. P1.11 Two-lens combination

1.12 Show Eqs. (1.4-26) and (1.4-27).

1.13 Show Eq. (1.4-22) by mathematical induction.

1.14 For the resonator shown in Fig. 1.19, draw the ray propagation diagram similar to that shown in Fig. 1.21 for the following parameters for $d > 0$:

a) $R_1 = \infty$ and $R_2 = -2d$ (hemispherical resonator);

b) $R_1 = d/2, R_2 = -d/2$ (concentric resonator);

c) $R_1 = d$ and $R_2 = \infty$;

d) $R_1 = R_2 = \infty$ (planar resonator) .

Bibliography

Banerjee, P.P. and T.-C. Poon (1991). *Principles of Applied Optics.* Irwin, Illinois.

Feynman, R., R. B. Leighton and M. Sands (1963). *The Feynman Lectures on Physics.* Addison-Wesley, Reading, Massachusetts.

Fowles, G. R. and G. L. Cassiday (2005). *Analytical Mechanics* (7th ed.). Thomson Brooks/Cole, Belmont, CA.

Gerard, A. and J. M. Burch (1975). *Introduction to Matrix Methods in Optics.* Wiley, New York.

Ghatak, A. K. (1980). *Optics.* Tata McGraw-Hill, New Delhi.

Goldstein, H. (1950). *Classical Mechanics.* Addison-Wesley, Reading, Massachusetts.

Hecht, E. and A. Zajac (1975). *Optics.* Addison-Wesley, Reading, Massachusetts.

Klein, M. V.(1970). *Optics.* Wiley, New York.

Nussbaum, A. and R. A. Phillips (1976). *Contemporary Optics for Scientists and Engineers,* Prentice-Hall, New York.

Pedrotti F. L. and L. S. Pedrotti (1987). *Introduction to Optics.* Prentice-Hall, Inc.,Englewood Cliffs, New Jersey.

Poon T.-C. and P. P. Banerjee (2001). *Contemporary Optical Image Processing with MATLAB* ® Elsevier, Oxford, UK.

Chapter 2

Wave Propagation and Wave Optics

In Chapter 1, we introduced some of the concepts of geometrical optics. However, geometrical optics cannot account for wave effects such as diffraction. In this Chapter, we introduce wave optics by starting from Maxwell's equations and deriving the wave equation. We thereafter discuss solutions of the wave equation and review power flow and polarization. We then discuss boundary conditions for electromagnetic fields and subsequently derive Fresnel's equations. We also discuss Fourier transform and convolution and then develop diffraction theory through the use of the Fresnel diffraction formula, which is derived in a unique manner using Fourier transforms. In the process, we define the spatial frequency transfer function and the spatial impulse response in Fourier optics. We also describe the distinguishing features of Fresnel and Fraunhofer diffraction and provide several illustrative examples. In the context of diffraction, we also develop wavefront transformation by lens and show the Fourier transforming properties of the lens. We also analyze resonators and the diffraction of a Gaussian beam. Finally, in the last Section of this chapter, we discuss Gaussian beam optics and introduce the q-transformation of Gaussian beams. In all cases, we restrict ourselves to wave propagation in a medium with a constant refractive index (homogenous medium). Beam propagation in inhomogeneous media is covered in Chapter 3.

2.1 Maxwell's Equations: A Review

In the study of electromagnetics, we are concerned with four vector quantities called electromagnetic (EM) fields: the electric field strength $\mathcal{E}$ (V/m), the electric flux density $\mathcal{D}$ (C/m^2), the magnetic field strength $\mathcal{H}$ (A/m), and the magnetic flux density $\mathcal{B}$ (Wb/m^2). The fundamental theory of electromagnetic fields is based on *Maxwell's equations*. In differential form, these are expressed as

$$\nabla \cdot \boldsymbol{D} = \rho_v, \tag{2.1-1}$$

$$\nabla \cdot \boldsymbol{B} = 0, \tag{2.1-2}$$

$$\nabla \times \boldsymbol{E} = -\frac{\partial \boldsymbol{B}}{\partial t}, \tag{2.1-3}$$

$$\nabla \times \boldsymbol{H} = \boldsymbol{J} + \frac{\partial \boldsymbol{D}}{\partial t}, \tag{2.1-4}$$

where $\boldsymbol{J}$ is the current density (A/m^2) and ρ_v denotes the electric charge density (C/m^3). $\boldsymbol{J}$ and ρ_v are the sources generating the electromagnetic fields. Maxwell's equations express the physical laws governing the *electric fields* $\boldsymbol{E}$ and $\boldsymbol{D}$, *magnetic fields* $\boldsymbol{H}$ and $\boldsymbol{B}$, and the sources $\boldsymbol{J}$ and ρ_v. From Eqs. (2.1-3) and (2.1-4), we see that a time-varying magnetic field produces a time-varying electric field and conversely, a time-varying electric field produces a time-varying magnetic field. It is precisely this coupling between the electric and magnetic fields generates electromagnetic waves capable of propagating through a medium even in free space. Note that, however, in the *static* case, none of the quantities in Maxwell's equations are a function of time. This happens when all charges are fixed in space or if they move in a steady rate such that ρ_v and $\boldsymbol{J}$ remain constant in time. Therefore, in the static case, Eqs. (2.1-3) and (2.1-4) becomes $\nabla \times \boldsymbol{E} = 0$ and $\nabla \times \boldsymbol{H} = \boldsymbol{J}$, respectively. The electric and magnetic fields are independent and not coupled together, which leads to the study of *electrostatics* and *magnetostatics*. Equation (2.1-1) is the differential representation of *Guass's law for electric fields*. To convert this to an integral form, which is more physically transparent, we integrate Eq. (2.1-1) over a volume v bounded by a surface S and use the *divergence theorem* (or *Gauss's theorem*),

$$\int_v \nabla \cdot \boldsymbol{D} dv = \oint_S \boldsymbol{D} \cdot \boldsymbol{ds}, \tag{2.1-5}$$

to get

$$\oint_S \boldsymbol{D} \cdot \boldsymbol{ds} = \int_v \rho_v dv. \tag{2.1-6}$$

This states that the *electric flux* $\oint_S \boldsymbol{D} \cdot \boldsymbol{ds}$ flowing out of a surface S enclosing a volume v equals the total charge enclosed in the volume. Equation (2.1-2) is *Gauss's law for magnetic fields*, which is the magnetic analog of Eq. (2.1-1) and can be converted to an integral form similar to Eq. (2.1-6) by using the divergence theorem once again,

$$\oint_S \boldsymbol{B} \cdot \boldsymbol{ds} = 0. \tag{2.1-7}$$

The right-hand sides of Eqs. (2.1-2) and (2.1-7) are zero because magnetic monopoles do not exist. Hence, the *magnetic flux* is always conserved.

Equation (2.1-3) enunciates *Faraday's law of induction*. To convert this to an integral form, we integrate over an open surface S bounded by a contour C and use *Stokes's theorem*,

$$\int_S (\nabla \times \boldsymbol{E}) \cdot \boldsymbol{ds} = \oint_S \boldsymbol{E} \cdot \boldsymbol{dl}, \tag{2.1-8}$$

to get

$$\oint_S \boldsymbol{E} \cdot \boldsymbol{dl} = -\int_S \frac{\partial \boldsymbol{B}}{\partial t} \cdot \boldsymbol{ds}. \tag{2.1-9}$$

This states that the *electromotive force* (EMF) $\oint_C \boldsymbol{E} \cdot \boldsymbol{dl}$ induced in a loop is equal to the time rate of change of the magnetic flux passing through the area of the loop. The EMF is induced in such a way that it opposes the variation of the magnetic field, as indicated by the minus sign in Eq. (2.1-9); this is known as *Lenz's law*. Similarly, the integral form of Eq. (2.1-4) reads

$$\oint_C \boldsymbol{\mathcal{H}} \cdot \boldsymbol{dl} = I = \int_S \frac{\partial \boldsymbol{D}}{\partial t} \cdot \boldsymbol{ds} + \int_S \boldsymbol{J} \cdot \boldsymbol{ds}, \tag{2.1-10}$$

which states that the line integral of $\boldsymbol{\mathcal{H}}$ around a closed loop C equals the total current I passing through the surface of the loop. When first formulated by Ampere, Eqs. (2.1-4) and (2.1-10) only had the current density term $\boldsymbol{J}$ on the right-hand side. Maxwell proposed the addition of the displacement current term $\partial \boldsymbol{D} / \partial t$ to include the effect of currents flowing through, for instance, a capacitor. For a given current and charge density distribution, we solve Maxwell's equations. However, we remark that Eq. (2.1-1) is not independent of Eq. (2.1-4) and, similarly, Eq. (2.1-2) is a consequence of Eq. (2.1-3). By taking the divergence on both sides of Eq. (2.1-4) and using the *continuity equation*:

$$\nabla \cdot \boldsymbol{J} + \frac{\partial \rho_v}{\partial t} = 0, \qquad (2.1\text{-}11)$$

which is the *principle of conservation of charge,* we can show that $\nabla \cdot \boldsymbol{D} = \rho_v$. Similarly, Eq. (2.1-2) is a consequence of Eq. (2.1-3). Hence, from Eqs.(2.1-1) to (2.1-4), we really have six independent scalar equations (three scalar equations for each curl equation) and twelve unknowns. The unknowns are the x,y and z components of $\boldsymbol{E}$, $\boldsymbol{D}$, $\boldsymbol{H}$, and $\boldsymbol{B}$. The six more scalar equations required are provided by the *constitutive relations*:

$$\boldsymbol{D} = \varepsilon \boldsymbol{E} \qquad (2.1\text{-}12a)$$

and

$$\boldsymbol{B} = \mu \boldsymbol{H}, \qquad (2.1\text{-}12b)$$

where ε denotes the permittivity (F/m) and μ the permeability (H/m) of the medium. In this book, we do not consider any magnetic media and hence μ is considered as a constant. ε becomes a function of position, i.e., $\varepsilon (x,y)$, when considering inhomogeneous media such as in fiber optics (Chapter 3). When it is also a function of time i.e., $\varepsilon(x,y,t)$, we consider acousto-optics (Chapter 4). It becomes a tensor when considering anisotropic media as in electro-optics (Chapter 5). However, in many cases, we can take ε and μ to be scalar constants. Indeed, this is true for a *linear, homogeneous,* and *isotropic* medium. A medium is *linear* if its properties do not depend on the amplitude of the fields in the medium. It is *homogeneous* if its properties are not functions of space. The medium is *isotropic* if its properties are the same in every direction from any given point.

　　　　Returning our focus to linear, homogeneous, and isotropic media, constants worth remembering are the values of ε and μ for free space or vacuum: $\varepsilon_0 = (1 / 36\pi) \times 10^{-9}$ F/m and $\mu_0 = 4\pi \times 10^{-7}$ H/m. For *dielectrics,* the value of ε is greater than ε_0, because the $\mathcal{D}$ field is composed of free-space part $\varepsilon_0\mathcal{E}$ and a material part characterized by a *dipole moment density* $\mathbf{p}$ (C/m^2). $\mathbf{p}$ is related to the electric field $\mathcal{E}$ as

$$\mathbf{p} = \chi\varepsilon_0\mathcal{E}, \qquad (2.1\text{-}13)$$

where χ is the *electric susceptibility* and indicates the ability of the electric dipoles in the dielectric to align themselves with the electric field. The $\mathcal{D}$ field is the sum of $\varepsilon_0\mathcal{E}$ and $\mathbf{p}$,

$$\mathcal{D} = \varepsilon_0\mathcal{E} + \mathbf{p} = \varepsilon_0\left(1 + \chi\right)\mathcal{E} = \varepsilon_0\varepsilon_r\mathcal{E}, \qquad (2.1\text{-}14)$$

where ε_r is the relative permittivity, so that

$$\varepsilon = \varepsilon_0(1 + \chi) = \varepsilon_0\varepsilon_r. \qquad (2.1\text{-}15)$$

Similarly, for magnetic materials $\mu = \mu_0\mu_r$ is greater than μ_0, where μ_r is the relative permeability. In free space, $\varepsilon_r = \mu_r = 1$.

2.2 Linear Wave Propagation

In this section, we first derive the wave equation and review some of the traveling-wave type solutions of the equation in different coordinate systems. We define the concept of *intrinsic impedance*, *Poynting vector* and *intensity*, and introduce the subject of *polarization*.

2.2.1 Traveling-wave solutions

In Section 2.1 we enunciated Maxwell's equations and the constitutive relations. For a given $\mathbf{J}$ and ρ_v, we remarked that we could, in fact, solve for the components of the electric field $\mathcal{E}$. In this subsection, we see how this can be done. We derive the wave equation describing the propagation of the electric and magnetic fields and find its general solutions in different coordinate systems. By taking the curl of both sides of Eq. (2.1-3), we have

$$\nabla \times \nabla \times \boldsymbol{E} = -\nabla \times \frac{\partial \boldsymbol{B}}{\partial t}$$

$$= -\frac{\partial}{\partial t}(\nabla \times \boldsymbol{B}) = -\mu \frac{\partial}{\partial t}(\nabla \times \boldsymbol{H}),$$

(2.2-1)

where we have used the second of the constitutive relations [Eq. (2.1-12b)] and assumed μ to be space- and time-independent. Now, employing Eq. (2.1-4), Eq. (2.2-1) becomes

$$\nabla \times \nabla \times \boldsymbol{E} = -\mu\varepsilon \frac{\partial^2 \boldsymbol{E}}{\partial t^2} - \mu \frac{\partial \boldsymbol{J}}{\partial t},$$

(2.2-2)

where we have used the first of the constitutive relations [Eq. (2.1-11a)] and assumed ε to be time-independent. Then, by using the following vector identity ($\boldsymbol{A}$ is some arbitrary vector)

$$\nabla \times \nabla \times \boldsymbol{A} = \nabla(\nabla \cdot \boldsymbol{A}) - \nabla^2 \boldsymbol{A}, \qquad \nabla^2 = \nabla \cdot \nabla,$$

(2.2-3)

in Eq. (2.2-2), we get

$$\nabla^2 \boldsymbol{E} - \mu\varepsilon \frac{\partial^2 \boldsymbol{E}}{\partial t^2} = \mu \frac{\partial \boldsymbol{J}}{\partial t} + \nabla(\nabla \cdot \boldsymbol{E}).$$

(2.2-4)

If we now assume the permittivity ε to be space-independent as well, then we can recast the first of Maxwell's equations [Eq. (2.1-1)] in the form

$$\nabla \cdot \boldsymbol{E} = \frac{\rho_v}{\varepsilon},$$

(2.2-5)

using the first of the constitutive relations [Eq. (2.1-12a)]. Incorporating Eq. (2.2-5) into Eq. (2.2-4), we finally obtain

$$\nabla^2 \boldsymbol{E} - \mu\varepsilon \frac{\partial^2 \boldsymbol{E}}{\partial t^2} = \mu \frac{\partial \boldsymbol{J}}{\partial t} + \frac{1}{\varepsilon} \nabla \rho_v,$$

(2.2-6)

which is a *wave equation* having source terms on the right-hand side. This is the wave equation for $\boldsymbol{E}$ in a linear, homogeneous, and isotropic medium characterized by μ and ε.

Equation (2.2-6) is equivalent to three scalar equations, one for every component of $\mathcal{E}$. Expressions for the Laplacian (∇^2) operator in Cartesian (x, y, z), cylindrical (r, ϕ, z), and spherical (R, θ, ϕ) coordinates are given as follows:

$$\nabla^2 = \frac{\partial^2}{\partial x^2} + \frac{\partial^2}{\partial y^2} + \frac{\partial^2}{\partial z^2}, \tag{2.2-7}$$

$$\nabla^2 = \frac{\partial^2}{\partial r^2} + \frac{1}{r}\frac{\partial}{\partial r} + \frac{1}{r^2}\frac{\partial^2}{\partial \phi^2} + \frac{\partial^2}{\partial z^2}; \tag{2.2-8}$$

$$\nabla^2 = \frac{\partial^2}{\partial R^2} + \frac{2}{R}\frac{\partial}{\partial R} + \frac{1}{R^2 \sin^2\theta}\frac{\partial^2}{\partial \phi^2} + \frac{1}{R^2}\frac{\partial^2}{\partial \theta^2} + \frac{\cot\theta}{R^2}\frac{\partial}{\partial \theta}. \tag{2.2-9}$$

In space free of all sources ($J = 0, \rho_v = 0$), Eq. (2.2-6) reduces to the *homogeneous wave equation*

$$\nabla^2\mathcal{E} = \mu\varepsilon \frac{\partial^2 \mathcal{E}}{\partial t^2}. \tag{2.2-10}$$

A similar equation may be derived for the magnetic field $\mathcal{H}$,

$$\nabla^2\mathcal{H} = \mu\varepsilon \frac{\partial^2 \mathcal{H}}{\partial t^2}. \tag{2.2-11}$$

We caution readers that the ∇^2 operator, as written in Eqs. (2.2-7)-(2.2-9), must be applied only after decomposing Eqs. (2.2-10) and (2.2-11) into scalar equations for three orthogonal components. However, for the rectangular coordinate case only, these scalar equations may be recombined and interpreted as the Laplacian ∇^2 acting on the total vector.

Note that the quantity $\mu\varepsilon$ has the units of $(1/velocity)^2$. We call this velocity u and define it as

$$u^2 = \frac{1}{\mu\varepsilon}. \tag{2.2-12}$$

For free space, $\mu = \mu_0$, $\varepsilon = \varepsilon_0$ and $u = c$. We can calculate the value of c from the values of ε_0 and μ_0 mentioned in Section 2.1. This works out to about 3×10^8 m/s. This theoretical value, first calculated by Maxwell, was in remarkable agreement with Fizeau's previously measured speed of light (315,300 km/s). This led Maxwell to conclude that light is an electromagnetic disturbance in the form of waves propagating through the medium according to electromagnetic laws. Let us now examine the solutions of equations of the type of Eqs. (2.2-10) or (2.2-11) in different coordinate systems. For simplicity, we will analyze the homogeneous wave equation

$$\frac{\partial^2 \psi}{\partial t^2} - u^2 \nabla^2 \psi = 0, \qquad (2.2\text{-}13)$$

where ψ may represent a component of the electric field $\boldsymbol{E}$ or of the magnetic field $\boldsymbol{\mathcal{H}}$ and where u is the velocity of the wave in the medium. In Cartesian coordinates, the general solution is

$$\begin{aligned} \psi(x,y,z,t) &= c_1 f\left(\omega_0 t - k_{0x} x - k_{0y} y - k_{0z} z\right) \\ &\quad + c_2 g\left(\omega_0 t + k_{0x} x + k_{0y} y + k_{0z} z\right) \end{aligned} \qquad (2.2\text{-}14)$$

with the condition

$$\frac{\omega_0^2}{k_{0x}^2 + k_{0y}^2 + k_{0z}^2} = \frac{\omega_0^2}{k_0^2} = u^2, \qquad (2.2\text{-}15)$$

where c_1 and c_2 are some constants. In Eq. (2.2-15), ω_0 is the *angular frequency* (rad/s) of the wave and k_0 is the *propagation constant* (rad/m) in the medium. Since the ratio ω_0 / k_0 is a constant, the medium of propagation is said to be *nondispersive*. We can then re-express Eq. (2.2-14) as

$$\psi(x,y,z,t) = c_1 f\left(\omega_0 t - \mathbf{k}_0 \cdot \mathbf{R}\right) + c_2 g\left(\omega_0 t + \mathbf{k}_0 \cdot \mathbf{R}\right), \qquad (2.2\text{-}16)$$

where

$$\mathbf{R} = x\mathbf{a}_x + y\mathbf{a}_y + z\mathbf{a}_z, \qquad (2.2\text{-}17a)$$

$$\mathbf{k}_0 = k_{0x}\mathbf{a}_x + k_{0y}\mathbf{a}_y + k_{0z}\mathbf{a}_z. \qquad (2.2\text{-}17b)$$

$\mathbf{k}_0$ is called the *propagation vector* and $|\mathbf{k}_0| = k_0$; $\mathbf{a}_x, \mathbf{a}_y$, and $\mathbf{a}_z$ denote the unit vectors in the x, y, and z directions, respectively.

In one spatial dimension, i.e., $\psi(z,t)$, the wave equation [Eq. (2.2-13)] becomes

$$\frac{\partial^2 \psi}{\partial t^2} - u^2 \frac{\partial^2 \psi}{\partial z^2} = 0 \qquad (2.2\text{-}18)$$

and its general solution is

$$\psi(z,t) = c_1 f(\omega_0 t - k_0 z) + c_2 g(\omega_0 t + k_0 z), \quad u = \frac{\omega_0}{k_0}. \qquad (2.2\text{-}19)$$

Note that Eq. (2.2-14) or (2.2-16) comprises the superposition of two waves, traveling in opposite directions. We can define a *wave* as a disturbance of some form characterized by a recognizable amplitude and a recognizable velocity or propagation. Let us now consider a special case: $c_1 = 1$, $c_2 = 0$, and $\psi(.)$ is of the functional form of $\exp[j(.)]$, we then have the following solution:

$$\psi(x,y,z,t) = \exp\left[j(\omega_0 t - \mathbf{k}_0 \cdot \mathbf{R})\right]. \qquad (2.2\text{-}20)$$

This is called a plane-wave solution and the wave is called a *plane wave* of unit amplitude as $c_1 = 1$.

Consider now the cylindrical coordinate system. The simplest case is that of cylindrical symmetry, which requires that $\psi(r,\phi,z,t) = \psi(r,z,t)$. The ϕ-independence means that a plane perpendicular to the z axis will intersect the wavefront in a circle. Even in this very simple case, no solutions in terms of arbitrary functions can be found as was done previously for plane waves. However, we can show that the exact solution has a *Bessel-function* type dependence on r if we assume ψ to be *time-harmonic*, that is, of the form $\psi(r,t) = \text{Re}[\psi_p(r)e^{j\omega_0 t}]$, where Re[.] means "the real part of the quantity being bracketed," and $\psi_p(r)$ is a phasor corresponding to the time-varying field $\psi(r,t)$. For $r \gg 0$, the solution becomes

$$\psi(r,t) \approx \frac{1}{\sqrt{r}} \exp\left[j(\omega_0 t \pm k_0 r)\right] \qquad (2.2\text{-}21)$$

approximately satisfying the wave equation [Eq. (2.2-13)]. This wave is commonly called a *cylindrical wave*.

Finally, we present solutions of the wave equation in a spherical coordinate system. For spherical symmetry $(\partial / \partial\phi = 0 = \partial / \partial\theta)$, the wave equation, Eq. (2.2-13), with Eq. (2.2-9) assumes the form

$$R\left(\frac{\partial^2\psi}{\partial R^2} + \frac{2}{R}\frac{\partial\psi}{\partial R}\right) = \frac{\partial^2(R\psi)}{\partial R^2} = \frac{1}{u^2}\frac{\partial^2(R\psi)}{\partial t^2}. \qquad (2.2\text{-}22)$$

Now, the above equation is of the same form as Eq. (2.2-18). Hence, using Eq. (2.2-19), we can write down the solution of Eq. (2.2-22) as

$$\psi = \frac{c_1}{R} f(\omega_0 t - k_0 R) + \frac{c_2}{R} g(\omega_0 t + k_0 R) \qquad (2.2\text{-}23)$$

with $\omega_0 / k_0 = u$. Again, for a special case: $c_1 = 1$, $c_2 = 0$, and $\psi(.)$ is of the functional form of $\exp[j(.)]$, Eq. (2.2-23) takes the form of

$$\psi(R,t) = \frac{1}{R}\exp[j(\omega_0 t - k_0 R)], \qquad (2.2\text{-}24)$$

which is called a *spherical wave.*

2.2.2 *Maxwell's equations in phasor domain: Intrinsic impedance, the Poynting vector, and polarization*

The electromagnetic fields and the current and charge densities considered so far are real functions of space and time. When we are concerned, as assumed in previous sections, with time-harmonic fields at a single frequency ω_0, we can define, for example, the electric field as

$$\boldsymbol{\mathcal{E}}(x,y,z,t) = \mathrm{Re}\left[\mathbf{E}(x,y,z)\exp(j\omega_0 t)\right], \qquad (2.2\text{-}25)$$

where $\mathbf{E}(x,y,z)$ is a *vector field phasor* corresponding to the time-varying electric field $\boldsymbol{\mathcal{E}}(x,y,z,t)$. The phasor is complex in general as it has amplitude and phase information. Similar definitions apply to the other fields and to $\boldsymbol{J}$ and ρ_v, i.e., we have the following phasors:

$$\mathbf{B}(x,y,z), \ \mathbf{D}(x,y,z), \mathbf{H}(x,y,z) \ , \mathbf{J}(x,y,z) \text{ and } \rho_v(x,y,z).$$

With these phasors for the time-harmonic quantities and for a linear, isotropic, and homogenous medium characterized by ε and μ, we can formulate Maxwell's equations as follows:

$$\nabla \cdot \mathbf{D} = \rho_v(x,y,z), \tag{2.2-26a}$$

$$\nabla \cdot \mathbf{B} = 0, \tag{2.2-26b}$$

$$\nabla \times \mathbf{E} = -j\omega_0 \mathbf{B}, \tag{2.2-26c}$$

$$\nabla \times \mathbf{H} = \mathbf{J} + j\omega_0 \mathbf{D} \tag{2.2-26d}$$

with $\mathbf{D} = \varepsilon \mathbf{E}$ and $\mathbf{B} = \mu \mathbf{H}$. In obtaining Eq.(2.2-26), we have used the fact that

$$\begin{aligned}\nabla \cdot \boldsymbol{g}(x,y,z,t) &= \nabla \cdot \mathrm{Re}[\mathbf{G}(x,y,z)\exp(j\omega_0 t)] \\ &= \mathrm{Re}[\nabla \cdot \mathbf{G}(x,y,z)\exp(j\omega_0 t)]\end{aligned} \tag{2.2-27a}$$

for divergence operations, and

$$\begin{aligned}\nabla \times \boldsymbol{g}(x,y,z,t) &= \nabla \times \mathrm{Re}[\mathbf{G}(x,y,z)\exp(j\omega_0 t)] \\ &= \mathrm{Re}[\nabla \times \mathbf{G}(x,y,z)\exp(j\omega_0 t)]\end{aligned} \tag{2.2-27b}$$

for curl operations, and finally

$$\begin{aligned}\frac{\partial \boldsymbol{g}(x,y,z,t)}{\partial t} &= \frac{\partial}{\partial t}\{\mathrm{Re}[\mathbf{G}(x,y,z)\exp(j\omega_0 t)]\} \\ &= \mathrm{Re}[j\omega_o \mathbf{G}(x,y,z)\exp(j\omega_0 t)]\end{aligned} \tag{2.2-27c}$$

for time derivative operations, where $\mathbf{G}(x,y,z)$ is the phasor corresponding to the time-harmonic field $\boldsymbol{g}(x,y,z,t)$.

In an unbounded isotropic, linear, homogenous medium free of sources, electromagnetic wave propagation is *transverse* in nature. This means that the only components of $\boldsymbol{E}$ and $\boldsymbol{\mathcal{H}}$ are those that are transverse to the direction of propagation. To check this, we consider a simple case where the propagating electric and magnetic fields are traveling along the positive z-direction:

$$\mathbf{E} = \mathbf{E_x} + \mathbf{E_y} + \mathbf{E_z}$$
$$= E_{0x}\exp(-jk_0z)\mathbf{a_x} + E_{0y}\exp(-jk_0z)\mathbf{a_y}$$
$$+ E_{0z}\exp(-jk_0z)\mathbf{a_z},\qquad\qquad(2.2\text{-}28a)$$

$$\mathbf{H} = \mathbf{H_x} + \mathbf{H_y} + \mathbf{H_z}$$
$$= H_{0x}\exp(-jk_0z)\mathbf{a_x} + H_{0y}\exp(-jk_0z)\mathbf{a_y}$$
$$+ H_{0z}\exp(-jk_0z)\mathbf{a_z},\qquad\qquad(2.2\text{-}28b)$$

where E_{0x}, E_{0y}, E_{0z} and H_{0x}, H_{0y}, H_{0z} are complex constants in general. Again, if we want to convert the phasor quantity back into the time domain, we simply use Eq. (2.2-25).

We put Eq. (2.2-28a) in the first of Maxwell's equations [i.e., Eq. (2.2-26a)], with $\rho_v = 0$, and invoke the constitutive relations to obtain

$$\frac{\partial}{\partial z}\{E_{0z}\exp(-jk_0z)]\} = -jk_0 E_{0z}\exp(-jk_0z) = 0$$

implying

$$E_{0z} = 0.\qquad\qquad(2.2\text{-}29a)$$

This means there is no component of the electric field in the direction of propagation. The only possible components of **E** then must be in a plane transverse to the direction of propagation. Similarly, using Eqs. (2.2- 26b), we can show that

$$H_{0z} = 0.\qquad\qquad(2.2\text{-}29b)$$

Furthermore, substitution of Eq. (2.2-28) with $E_{0z} = 0 = H_{0z}$ into the third of Maxwell's equations, Eq. (2.2-26c), yields

$$k_0 E_{0y}\mathbf{a_x} - k_0 E_{0x}\mathbf{a_y} = -\mu\omega_0(H_{0x}\mathbf{a_x} + H_{0y}\mathbf{a_y}).$$

We can then write [using Eqs. (2.2-12) and (2.2-15)]

$$H_{0x} = -\frac{1}{\eta}E_{0y},\ H_{0y} = \frac{1}{\eta}E_{0x},\qquad\qquad(2.2\text{-}30)$$

where

$$\eta = \frac{\omega_0}{k_0}\mu = u\mu = \left(\frac{\mu}{\varepsilon}\right)^{1/2}$$

is called the *intrinsic* or *characteristic impedance* of the medium. The characteristic impedance has the units of V/A, or ohms [Ω]. Its value for free space is $\eta_0 = 377\ \Omega$. Now using Eqs. (2.2-28)-(2.2-30), we can show that

$$\mathbf{E} \cdot \mathbf{H} = 0, \tag{2.2-31}$$

which means that the electric and magnetic fields are orthogonal to each other, and that $\mathbf{E} \times \mathbf{H}$ is along the direction of propagation (z) of the electromagnetic field. Similar relationships can be established in other coordinate systems.

Note that $\mathbf{E} \times \mathbf{H}$ has the units of W/m^2, reminiscent of power per unit area. All electromagnetic waves carry energy, and for isotropic media the energy flow occurs in the direction of propagation of the wave. However, this is not true for anisotropic media [see Chapter 5]. The *Poynting vector* S , defined as

$$S = \mathcal{E} \times \mathcal{H}, \tag{2.2-32}$$

is a power density vector associated with an electromagnetic field. In a linear, homogeneous, isotropic unbounded medium, we can consider a simple case and choose the electric and magnetic fields to be of the form

$$\mathcal{E}(z,t) = \text{Re}[E_0 \exp[j(\omega_0 t - k_0 z)]]\mathbf{a}_x, \tag{2.2-33a}$$

and

$$\mathcal{H}(z,t) = \text{Re}\left[\frac{E_0}{\eta}\exp[j(\omega_0 t - k_0 z)]\right]\mathbf{a}_y, \tag{2.2-33b}$$

where E_0 is, in general, a complex quantity. This choice is consistent with Eqs. (2.2-30)-(2.2-31). Note that S is a function of time and it is more convenient, therefore, to define the time-averaged power density $<S>$ as

$$< S >= \frac{\omega_o}{2\pi} \int_0^{2\pi/\omega_o} S dt = \frac{|E_0|^2}{2\eta} \mathbf{a_z} = \varepsilon u \frac{|E_0|^2}{2} \mathbf{a_z} . \qquad (2.2\text{-}34)$$

We can also find the time-averaged power density by conveniently using phasors instead of the time integration:

$$< S >= \frac{1}{2} \text{Re}[\mathbf{E} \times \mathbf{H}^*]. \qquad (2.2\text{-}35)$$

We can now define *irradiance I* as the magnitude of $< S >$

$$I = |<S>|, \qquad (2.2\text{-}36)$$

which is often referred to as the *intensity*. However, the intensity is commonly taken to be proportional to the magnitude squared of the complex field, i.e., $|E_0|^2$.

We shall now introduce readers to the concept of *polarization* of the electric field. The polarization describes the locus of the tip of the $\boldsymbol{E}$ vector at a given point in space as time advances. A separate description of the magnetic field is not necessary because the direction of $\mathcal{H}$ is definitely related to that of $\boldsymbol{E}$.

Let us consider a plane wave propagating in the $+z$ direction and therefore the electric field is oriented in the $x - y$ plane, i.e., $E_{0z} = 0$, which has two components. According to Eq. (2.2-28a), we write

$$\mathbf{E}(z) = E_{0x} \exp(-jk_0 z)\mathbf{a_x} + E_{0y} \exp(-jk_0 z)\mathbf{a_y}, \qquad (2.2\text{-}37a)$$

where

$$E_{0x} = |E_{0x}|, E_{0y} = |E_{0y}| e^{-j\phi_0}, \qquad (2.2\text{-}37b)$$

with ϕ_0 being a constant.

First, we consider the case where $\phi_0 = 0$ or $\pm\pi$. The two components of $\mathbf{E}(z)$ are in phase, and

$$\mathcal{E}(z,t) = \text{Re}[\mathbf{E}(z)\exp(j\omega_0 t)]$$
$$= \left(\left|E_{0x}\right|\mathbf{a_x} \pm \left|E_{0y}\right|\mathbf{a_y}\right)\cos(\omega_0 t - k_0 z). \qquad (2.2\text{-}38)$$

The direction of $\mathcal{E}$ is fixed on a plane perpendicular to the direction of propagation (this plane is referred to as the *plane of polarization*) and does not vary with time, and the electric field is said to be *linearly polarized*. The plus sign in Eq. (2.2-38) refers to the case $\phi_0 = 0$, whereas the minus sign to the case $\phi_0 = \pm\pi$.

As a second case, we assume $\phi_0 = \pm\pi/2$ and $\left|E_{0x}\right| = \left|E_{0y}\right| = E_0$. In this case,

$$\mathcal{E}(z,t) = E_0 \cos(\omega_0 t - k_0 z)\mathbf{a_x} \pm E_0 \sin(\omega_0 t - k_0 z)\mathbf{a_y}. \qquad (2.2\text{-}39)$$

When monitored at a certain point $z = z_0$ during propagation, the direction of $\mathcal{E}$ is not longer fixed along a line, but varies with time according to $\theta(t) = \omega_0 t - k_0 z_0$, where θ represents the angle between $\mathcal{E}$ and the (transverse) x axis. The amplitude of $\mathcal{E}$ (which is equal to E_0) is, however, still a constant. This is an example of *circular polarization* of the electric field. When $\phi_0 = -\pi/2$, the y-component of $\mathcal{E}(z,t)$ leads the x-component of $\mathcal{E}(z,t)$ by $\pi/2$. Hence, as a function of time, $\mathcal{E}(z,t)$ describes a clockwise circle in the $x-y$ plane as seen head-on at $z = z_0$; therefore, we have the so-called *clockwise circularly polarized light* or *left-circularly polarized light*. Similarly, for $\phi_0 = +\pi/2$, $\mathcal{E}(z,t)$ describes a counter-clockwise circle, and we have *counter-clockwise circularly polarized light* or *right-circularly polarized light*.

In the general case when ϕ_0 is arbitrary,

$$\mathcal{E}(z,t) = \left|E_{0x}\right|\cos(\omega_0 t - k_0 z)\mathbf{a_x} + \left|E_{0y}\right|\cos(\omega_0 t - k_0 z - \phi_0)\mathbf{a_y}.$$
$$(2.2\text{-}40)$$

As in the case of circularly polarized waves (where $\left|E_{0x}\right|^2 + \left|E_{0y}\right|^2 = E_0^2 = $ constant), the direction of $\mathcal{E}(z,t)$ is no longer fixed and its tip traces an ellipse in the $x-y$ plane. The wave is said to be *elliptically polarized*.

Note that for values of ϕ_0 equal to 0 or $\pm\pi$ and $\pm\pi/2$ (with $|E_{0x}| = |E_{0y}|$ = E_0), the polarization configurations reduce to the linearly and circularly polarized cases, respectively.

   Figure 2.1 illustrates various polarization configurations corresponding to different values of ϕ_0 to demonstrate clearly linear, circular, and elliptical polarizations. In this figure, we show the direction of rotation of the $\boldsymbol{\mathcal{E}}$ field with time, and its magnitude for various ϕ_0. When $\phi_0 = 0$ or $\pm\pi$, the $\boldsymbol{\mathcal{E}}$ field is linearly polarized and the $\boldsymbol{\mathcal{E}}$ vector does not rotate but fixed along a line.

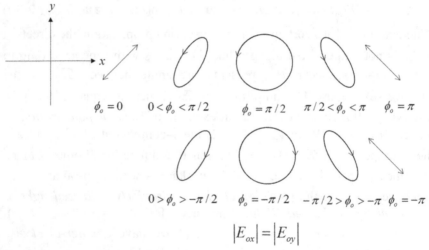

$\phi_o = 0$ $0 < \phi_o < \pi/2$ $\phi_o = \pi/2$ $\pi/2 < \phi_o < \pi$ $\phi_o = \pi$

$0 > \phi_o > -\pi/2$ $\phi_o = -\pi/2$ $-\pi/2 > \phi_o > -\pi$ $\phi_o = -\pi$

$$|E_{ox}| = |E_{oy}|$$

Fig. 2.1 Various polarization configurations corresponding to different values of ϕ_0 $(|E_{0x}| \neq |E_{0y}|)$, unless otherwise stated. For circular polarization, $(|E_{0x}| = |E_{0y}|)$.

2.2.3 *Electromagnetic waves at a boundary and Fresnel's equations*

We have studied wave propagation through an unbounded medium so far. In this section, we discuss wave propagation through a boundary between two semi-infinite media sharing a common interface. Specifically, we first investigate the effects of wave polarization upon reflection and transmission at the interface between two linear, isotropic and homogeneous media and derive the *Fresnel's equations.* We then include a discussion on *total internal reflection* and establish the properties of *evanescent waves.*

We consider a plane polarized wave incident on the interface at an angle θ_i with respect to the normal of the interface as shown in Fig. 2.2. The plane containing the incident propagation vector $\mathbf{k}_i$ and the normal to the interface is called the *plane of incidence.* Since a vector field lying on a plane in an arbitrary direction always can be decomposed into two orthogonal directions, we choose to decompose the $\mathcal{E}$ field into a direction perpendicular and the other parallel to the plane of incidence. We consider these two cases separately, and the general situation is obtained by superposing the results of the two cases.

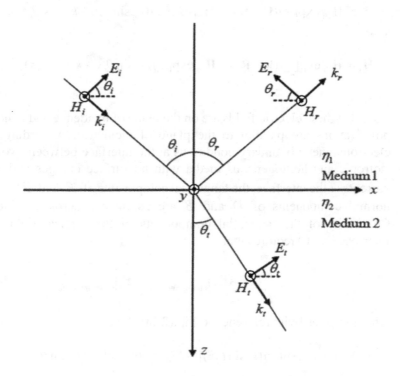

Fig. 2.2 Parallel polarization.

Parallel polarization

With reference to Fig. 2.2, we take the fields of the incident, reflected, and transmitted waves to be of the following forms, respectively

$$\mathbf{E}_i = \mathbf{E}_{i0} \exp[-j(\mathbf{k}_i \cdot \mathbf{R})] = \mathbf{E}_{i0} \exp[-j(k_i \sin\theta_i x + k_i \cos\theta_i z)],$$
$$\mathbf{E}_r = \mathbf{E}_{r0} \exp[-j(\mathbf{k}_r \cdot \mathbf{R})] = \mathbf{E}_{r0} \exp[-j(k_r \sin\theta_r x - k_r \cos\theta_r z)],$$
and

$$\mathbf{E}_t = \mathbf{E}_{t0} \exp[-j(\mathbf{k}_t \cdot \mathbf{R})] = \mathbf{E}_{t0} \exp[-j(k_t \sin\theta_t x + k_t \cos\theta_t z)],$$

$$(2.2\text{-}41a)$$

Similarly, for the magnetic fields, we have

$$\mathbf{H}_i = \mathbf{H}_{i0} \exp[-j(\mathbf{k}_i \cdot \mathbf{R})] = \mathbf{H}_{i0} \exp[-j(k_i \sin\theta_i x + k_i \cos\theta_i z)],$$
$$\mathbf{H}_r = \mathbf{H}_{r0} \exp[-j(\mathbf{k}_r \cdot \mathbf{R})] = \mathbf{H}_{r0} \exp[-j(k_r \sin\theta_r x - k_r \cos\theta_r z)],$$
and

$$\mathbf{H}_t = \mathbf{H}_{t0} \exp[-j(\mathbf{k}_t \cdot \mathbf{R})] = \mathbf{H}_{t0} \exp[-j(k_t \sin\theta_t x + k_t \cos\theta_t z)].$$

$$(2.2\text{-}41b)$$

Note that these electric fields are on the plane of incidence and hence their polarizations are parallel to the plane of incidence. According to the electromagnetic boundary conditions at the interface between two linear, isotropic and homogeneous media with no surface charges and surface currents at the interface, the tangential components of $\mathcal{E}$ and $\mathcal{H}$, and the normal components of $\mathcal{D}$ and $\mathcal{B}$ are continuous across an interface. Continuity of the tangential components of the electric fields at the interface ($z = 0$) requires that

$$(\mathbf{E_i} + \mathbf{E_r})\big|_{\text{along interface}} = \mathbf{E_t}\big|_{\text{along interface}}, \qquad (2\text{-}2\text{-}42a)$$

which implies [with reference to Fig.(2.2)]

$$E_{i0} \cos\theta_i \exp[-j(k_i \sin\theta_i x)] - E_{r0} \cos\theta_r \exp[-j(k_r \sin\theta_r x)]$$

$$= E_{t0} \cos \theta_t \exp[-j(k_t \sin \theta_t x)], \tag{2.2-42b}$$

where, according to Fig. 2.2, we have used

$$\mathbf{E}_{i0} = E_{i0}(\cos \theta_i \mathbf{a_x} - \sin \theta_i \mathbf{a_z}),$$
$$\mathbf{E}_{r0} = E_{r0}(-\cos \theta_r \mathbf{a_x} - \sin \theta_r \mathbf{a_z}),$$

and

$$\mathbf{E}_{t0} = E_{t0}(\cos \theta_t \mathbf{a_x} - \sin \theta_t \mathbf{a_z}).$$

Now, the boundary condition for the tangential component of the magnetic field gives

$$(\mathbf{H_i} + \mathbf{H_r})\big|_{\text{along interface}} = \mathbf{H_t}\big|_{\text{along interface}}, \tag{2.2-43a}$$

which is equivalent to

$$H_{i0} \exp[-j(k_i \sin \theta_i x)] + H_{r0} \exp[-j(k_r \sin \theta_r x)]$$
$$= H_{t0} \exp[-j(k_t \sin \theta_t x)], \tag{2.2-44}$$

where $\mathbf{H}_{i0} = H_{i0}\mathbf{a_y}$, $\mathbf{H}_{r0} = H_{r0}\mathbf{a_y}$, and $\mathbf{H}_{t0} = H_{t0}\mathbf{a_y}$. Now, in order to satisfy Eqs. (2.2-42b) and (2.2-43b) for all possible values of x along the interface, all three exponential arguments must be equal and that gives the so-called *phase matching condition*:

$$k_i \sin \theta_i = k_r \sin \theta_r = k_t \sin \theta_t. \tag{2.2-44}$$

Note that the first equality in the above equation leads to law of reflection and the second equality leads to Snell's law. In light of Eq. (2.2-44), the boundary conditions given by Eqs. (2.2-42b) and (2.2-43b) reduce to

$$E_{i0} \cos \theta_i - E_{r0} \cos \theta_r = E_{t0} \cos \theta_t, \tag{2.2-45a}$$

and

$$H_{i0} + H_{r0} = H_{t0}, \tag{2.2-45b}$$

respectively. Using $H_{i0} = E_{i0}/\eta_1, H_{r0} = E_{r0}/\eta_1$ and $H_{t0} = E_{t0}/\eta_2$, where η_1 and η_2 are the intrinsic impedances for medium 1 and 2, Eqs. (2.2-45a) and (2.2-45b) can be solved simultaneously to obtain the expressions for the *amplitude reflection and transmission coefficients*, $r_\parallel$ and $r_\perp$, respectively:

$$r_\parallel = \frac{E_{r0}}{E_{i0}} = \frac{\eta_1 \cos\theta_i - \eta_2 \cos\theta_t}{\eta_1 \cos\theta_i + \eta_2 \cos\theta_t}, \qquad (2.2\text{-}46a)$$

$$t_\parallel = \frac{E_{t0}}{E_{i0}} = \frac{2\eta_2 \cos\theta_i}{\eta_1 \cos\theta_i + \eta_2 \cos\theta_t}. \qquad (2.2\text{-}46b)$$

Perpendicular polarization

In this case, the electric field vectors are perpendicular to the plane of incidence, as shown in Fig. 2.3. As we did previously in the parallel-polarization case, we can obtain the following expressions for the amplitude reflection and transmission coefficients:

$$r_\perp = \frac{E_{r0}}{E_{i0}} = \frac{\eta_2 \cos\theta_i - \eta_1 \cos\theta_t}{\eta_2 \cos\theta_i + \eta_1 \cos\theta_t}, \qquad (2.2\text{-}47a)$$

$$t_\perp = \frac{E_{t0}}{E_{i0}} = \frac{2\eta_2 \cos\theta_i}{\eta_2 \cos\theta_i + \eta_1 \cos\theta_t}. \qquad (2.2\text{-}47b)$$

Equations (2.2-46) and (2.2-47) are called the *Fresnel's equations*, which dictate plane wave reflection and transmission at the interface between two semi-infinite media characterized by η_1 and η_2.

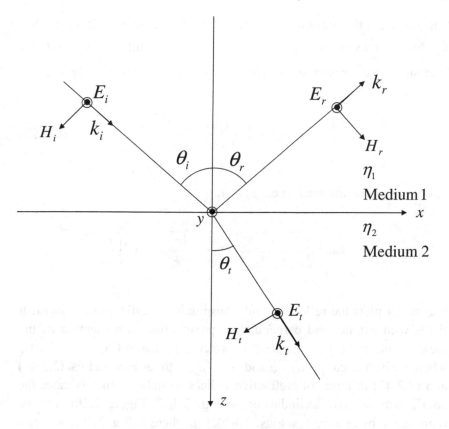

Fig. 2.3 Perpendicular polarization.

Brewster angle

The incident angle for which the reflection coefficient is zero is called the *Brewster angle* θ_p, also called the *polarizing angle*. For perpendicular polarization, we set $r_\perp = 0$ to get

$$\eta_2 \cos \theta_i = \eta_1 \cos \theta_t. \qquad (2.2\text{-}48)$$

Using Eqs. (2.2-30) and (2.2-44), we solve for θ_i in Eq. (2.2-48) to obtain

$$\sin \theta_i = \sqrt{\frac{1 - (\mu_1 \varepsilon_2 / \mu_2 \varepsilon_1)}{1 - (\mu_1 / \mu_2)^2}} = \sin \theta_{p\perp}. \qquad (2.2\text{-}49)$$

When $\mu_1 = \mu_2$, the denominator of Eq. (2.2-48) goes to zero. It means that θ_p does not exist for nonmagnetic materials. Similarly, we find the Brewster angle for parallel polarization by setting $r_{||} = 0$ to obtain

$$\sin\theta_{p||} = \sqrt{\frac{1-(\mu_2\varepsilon_1 / \mu_1\varepsilon_2)}{1-(\varepsilon_1 / \varepsilon_2)^2}}. \qquad (2.2\text{-}50)$$

For nonmagnetic materials, i.e., $\mu_1 = \mu_2$,

$$\theta_{p||} = \sin^{-1}\sqrt{\frac{1}{1+(\varepsilon_1 / \varepsilon_2)}} = \tan^{-1}\sqrt{\frac{\varepsilon_2}{\varepsilon_1}} = \tan^{-1}\left(\frac{n_2}{n_1}\right). \qquad (2.2\text{-}51)$$

Figure 2.4 plots the reflection and transmission coefficients for incident fields with parallel and perpendicular polarization as a function of the incident angle θ_i for an air-glass interface ($n_1 = 1, n_2 = 1.5$, $\mu_1 = \mu_2 = 1$), where we have used $\eta = \sqrt{\mu / \varepsilon}$ and $n = \sqrt{\mu_r\varepsilon_r}$ to re- express Eqs. (2.2-46) and (2.2-47) in terms of refractive indices n_1 and n_2. In this case, the coefficients are real as indicated in Fig. 2.4(a). Figure 2.4(a) can be represented by two figures, Figs. 2.4(b), (c), where in Fig. 2.4(c) we see a phase jump from zero degree to -180 degrees at the Brewster angle for the $r_{pa}(r_{||})$ curve.

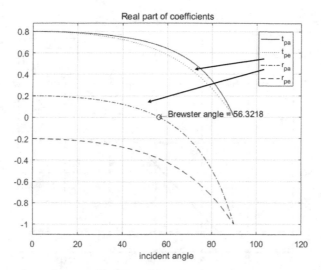

(a) $r_\parallel, t_\parallel, r_\perp, t_\perp$ vs. incident angle. Coefficients are all real in this case.

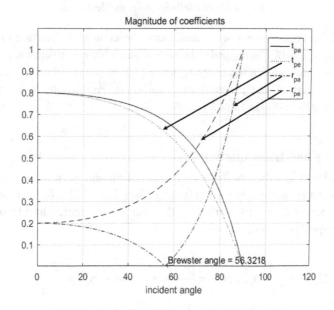

(b) Magnitude of coefficients vs. incident angle.

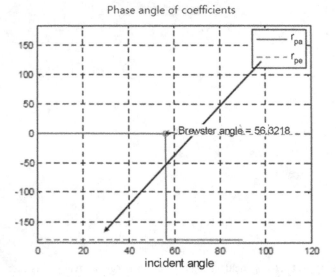

(c) Phase angle of coefficients vs. incident angle.

Fig. 2.4 Reflection and transmission coefficients for an air-glass interface ($n_1 = 1.0$, $n_2 = 1.5$): t_{pa} and t_{pe} correspond to the cases of transmission coefficient of parallel and perpendicular polarization, respectively. r_{pa} and r_{pe} corresponds to the cases of reflection coefficient of parallel and perpendicular polarization, respectively. These plots are generated using the m-file in Table 2.1.

Reflectivity and transmissivity

It is useful to relate the coefficients of reflection and of transmission to the flow of energy across the interface. Using Eq. (2.2-34), we write the averaged power densities carried by the incident, reflected and transmitted beams, respectively as follows:

$$<\mathbf{S}_i> = \frac{|E_{i0}|^2}{2\eta_1}\mathbf{a}_i, <\mathbf{S}_r> = \frac{|E_{r0}|^2}{2\eta_1}\mathbf{a}_r, \text{ and } <\mathbf{S}_t> = \frac{|E_{t0}|^2}{2\eta_2}\mathbf{a}_t,$$

where $\mathbf{a}_i, \mathbf{a}_r$ and $\mathbf{a}_t$ are the unit vectors of $\mathbf{k}_i, \mathbf{k}_r$ and $\mathbf{k}_t$, respectively as shown in Figs. 2.2 and 2.3. The coefficients of reflection R and transmission T are defined as the ratios of the average power across the interface. For R, it is given by

$$R = \frac{|<\mathbf{S_r}>\cdot\mathbf{a_z}|}{|<\mathbf{S_i}>\cdot\mathbf{a_z}|} = \frac{|E_{r0}|^2 \cos\theta_r}{|E_{i0}|^2 \cos\theta_i}. \tag{2.2-52}$$

Hence, for parallel and perpendicular polarization, we have

$$R_{\|} = |r_{\|}|^2 \text{ and } R_{\perp} = |r_{\perp}|^2, \tag{2.2-53}$$

respectively as $\theta_i = \theta_r$. $R_{\|}$ and $R_{\perp}$ are also called *reflectivity* or *reflectance* in optics. The coefficient of transmission (also called *transmissivity* or *transmittance* in optics) is

$$T = \frac{|<\mathbf{S_t}>\cdot\mathbf{a_z}|}{|<\mathbf{S_i}>\cdot\mathbf{a_z}|} = \frac{|E_{t0}|^2 \, \eta_1 \cos\theta_t}{|E_{i0}|^2 \, \eta_2 \cos\theta_i}. \tag{2.2-54}$$

Hence, for parallel and perpendicular polarization, we have

$$T_{\|} = \frac{\eta_1 \cos\theta_t}{\eta_2 \cos\theta_i}|t_{\|}|^2 \text{ and } T_{\perp} = \frac{\eta_1 \cos\theta_t}{\eta_2 \cos\theta_i}|t_{\perp}|^2 \tag{2.2-55}$$

Note that conservation of power requires that

$$R_{\perp} + T_{\perp} = 1 \text{ and } R_{\|} + T_{\|} = 1. \tag{2.2-56}$$

As a practical example, for normal incidence ($\theta_i = \theta_t = 0$) from air ($n_1 = 1$) to glass ($n_2 = 1.5$),

$$R_{\|} = R_{\perp} = |r_{\|}|^2 = |r_{\perp}|^2 = [(n_1 - n_2)/(n_1 + n_2)]^2 = 0.04,$$

and $T_{\perp} = T_{\|} = 0.96$. Hence, about 4% of the light is reflected and 96% is transmitted into glass.

Total internal reflection

Recall from Section 1.2 that for $n_1 > n_2$, any light ray incident at an angle greater than the critical angle, $\theta_c = \sin^{-1}(n_2 / n_1)$, experiences total internal reflection. What is the picture in terms of wave theory? It turns out Fresnel's equations are all applicable to total reflection if we disregard the fact that $\sin \theta_t > 1$ and for $\theta_i > \theta_c$, we set

$$\cos \theta_t = -(1 - \sin^2 \theta_t)^{\frac{1}{2}}$$

$$= -[1 - \left(\frac{n_1}{n_2}\right)^2 \sin^2 \theta_i]^{\frac{1}{2}} \qquad (2.2\text{-}57)$$

$$= \pm j[\left(\frac{n_1}{n_2}\right)^2 \sin^2 \theta_i - 1]^{\frac{1}{2}}.$$

Hence, from Eq. (2.2-41). We have, for reflected field,

$$\mathbf{E_r} = \mathbf{E_{r0}} \exp[-j(\mathbf{k_r} \cdot \mathbf{R})]$$
$$= \mathbf{E_{r0}} \exp[-j(k_r \sin \theta_r x - k_r \cos \theta_r z)],$$

and the transmitted field is

$$\mathbf{E_t} = \mathbf{E_{t0}} \exp[-j(\mathbf{k_t} \cdot \mathbf{R})]$$
$$= \mathbf{E_{t0}} \exp[-j(k_t \sin \theta_t x + k_t \cos \theta_t z)]$$

$$= \mathbf{E_{t0}} \exp\left[-jk_t\left(\frac{n_1}{n_2}\right)\sin \theta_i x\right] \exp\left\{-k_t\left[\left(\frac{n_1}{n_2}\right)^2 \sin^2 \theta_i - 1\right]^{\frac{1}{2}} z\right\},$$

where we have only retained the real exponential in z with a negative argument to prevent nonphysical solutions. We see that the transmitted field is propagating along the x-direction, with an exponentially decaying amplitude in the z-direction. Such a wave is called an *evanescent wave*.

Taking the case that the incident wave is polarized with its electrical field perpendicular to the plane of incidence [see Fig. 2.3], from Fresnel's equations (2.2-47) and the fact that $\cos \theta_t$ is now an imaginary quantity, $r_\perp$ and $t_\perp$ become complex and we find

$$r_\perp = |r_\perp| \exp(j\alpha) = \exp(j\alpha) \qquad (2.2\text{-}58)$$

and

$$t_\perp = |t_\perp| \exp(j\alpha/2) = \frac{2\cos\theta_i}{\sqrt{1-(n_2/n_1)^2}} \exp(j\alpha/2), \quad (2.2\text{-}59)$$

where

$$\alpha = 2\tan^{-1}\left(\frac{\sqrt{\sin^2\theta_i - (n_2/n_1)^2}}{\cos\theta_i}\right)$$

is the phase angle of the reflection coefficient. We can now write, assuming $\mathbf{E}_{i0} = E_{i0}\mathbf{a_y}$,

$$\mathbf{E_r} = \mathbf{E_{r0}}\exp[-j(\mathbf{k_r}\cdot\mathbf{R})] = E_{i0}\mathbf{a_y}\exp(j\alpha)\exp[-j(\mathbf{k_r}\cdot\mathbf{R})]. \quad (2.2\text{-}60)$$

and

$$\mathbf{E_t} = \mathbf{E_{t0}}\exp[-j(\mathbf{k_t}\cdot\mathbf{R})]$$

$$= E_{i0}\mathbf{a_y}|t_\perp|\exp(j\alpha)\exp[-jk_t(\frac{n_1}{n_2})\sin\theta_i x]$$

$$\times\exp\{-k_t[(\frac{n_1}{n_2})^2\sin^2\theta_i - 1]^{\frac{1}{2}}z\}. \quad (2.2\text{-}61)$$

We notice that the amplitude of the reflected wave is equal to that of the incidence and hence the energy is totally reflected [see Fig. 2.5(a)]. However, there is a phase change upon reflection, which varies from $0°$ at the critical angle to $180°$ at grazing incidence as illustrated in Fig. 2.5(b). The results of Fig. 2.5 are plotted using the m-file shown in Table 2.2. Now, the corresponding magnetic field for the transmitted field is, with reference to Fig. 2.3 and using Eq. (2.2-57),

$$\mathbf{H_t} = \mathbf{H_{t0}}\exp[-j(\mathbf{k_t}\cdot\mathbf{R})]$$

$$= (-\cos\theta_t\mathbf{a_x} + \sin\theta_t\mathbf{a_z})\frac{E_{i0}|t_\perp|}{\eta_2}\exp(j\alpha)\exp\left[-jk_t\left(\frac{n_1}{n_2}\right)\sin\theta_i x\right]$$

$$\times\exp\left\{-k_t\left[\left(\frac{n_1}{n_2}\right)^2\sin^2\theta_i - 1\right]^{\frac{1}{2}}z\right\},$$

and it becomes the following equation when we use Eq. (2.2-57):

$$\mathbf{H}_t = \left\{ j \left[\left(\frac{n_1}{n_2} \right)^2 \sin^2 \theta_1 - 1 \right]^{\frac{1}{2}} \mathbf{a_x} + \left(\frac{n_1}{n_2} \right) \sin \theta_i \mathbf{a_z} \right\} \frac{E_{i0} |t_\perp|}{\eta_2} \exp[j\alpha]$$

$$\times \exp \left[-jk_t \left(\frac{n_1}{n_2} \right) \sin \theta_i x \right] \exp \left\{ -k_t \left[\left(\frac{n_1}{n_2} \right)^2 \sin^2 \theta_i - 1 \right]^{\frac{1}{2}} z \right\}.$$

$$(2\text{-}2\text{-}62)$$

The time-averaged power density $< \mathbf{S_t} >$ of the transmitted field is then

$$< \mathbf{S_t} > \propto \mathrm{Re}[\mathbf{E_t} \times \mathbf{H_t^*}]$$

$$= \mathrm{Re} \left\{ \left[-j \left[\left(\frac{n_1}{n_2} \right)^2 \sin^2 \theta_i - 1 \right]^{\frac{1}{2}} \mathbf{a_z} + \left(\frac{n_1}{n_2} \right) \sin \theta_i \mathbf{a_x} \right] \right.$$

$$\left. \times \frac{E_{i0}^2 |t_\perp|^2}{\eta_2} \exp \left\{ -2k_t \left[\left(\frac{n_1}{n_2} \right)^2 \sin^2 \theta_i - 1 \right]^{\frac{1}{2}} z \right\} \right\}. \qquad (2\text{-}2\text{-}63)$$

Note that the transmitted field is obviously not zero $(|t_\perp| \neq 0)$, despite the fact that there is no power flowing along the z-direction (as the Poynting vector in th z-direction is imaginary). However, there is power flowing along the interface inside the less denser medium.

   If we consider a collection of plane waves (such as a beam), traveling in different directions, to be incident on the interface at angles larger than the critical angle, each plane wave experiences total internal reflection, and the reflection coefficient for each is different. Upon reflection, we can reconstruct the reflected beam by adding the complex amplitudes of each reflected plane wave. The net result is a reflected beam that is laterally shifted along the interface upon reflection. This lateral shift can be interpreted as the energy of the beam entering the less denser medium, traveling along the interface within the less denser medium, and then re-emerging from the less denser medium to the denser medium upon reflection. This lateral shift along the interface of the beam is known as the *Goos-Hänchen shift*.

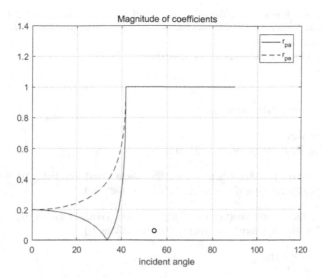

a) Magnitude of reflection coefficients vs. incident angle.

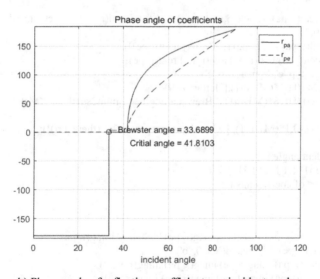

b) Phase angle of reflection coefficients vs. incident angle.

Fig. 2.5 Reflection coefficients for an glass-air interface $(n_1 = 1.5, n_2 = 1)$: r_{pa} and r_{pe} correspond to the cases of reflection coefficient of parallel and perpendicular polarization, respectively.

Table 2.1 Fresnel_Eq.m (m-file for plotting Fresnel's Eqs. (2.2-46) and (2.2-47)).

```
%Fresnel_Eq.m
%This m-file plots Fresnel equations (2.2-46) and (2.2-47)
n1=input('n1 =');
n2=input('n2 =');

theta_i=0:0.001:pi/2;%Incidence angle

z=(n1/n2)*sin(theta_i);
theta_t=-j*log(j*z+(ones(size(z))-z.^2).^0.5);

r_pa=(n2*cos(theta_i)-n1*cos(theta_t))./(n2*cos(theta_i)+n1*cos(theta_t));
t_pa=(2*n1*cos(theta_i))./(n2*cos(theta_i)+n1*cos(theta_t));

r_pe=(n1*cos(theta_i)-n2*cos(theta_t))./(n1*cos(theta_i)+n2*cos(theta_t));
t_pe=(2*n1*cos(theta_i))./(n1*cos(theta_i)+n2*cos(theta_t));

[N M]=min(abs(r_pa));
theta_i=theta_i/pi*180;
theta_cri=asin(n2/n1)*180/pi;

figure(1)
plot(theta_i,real(t_pa),'-',theta_i,real(t_pe),':',theta_i,real(r_pa),'-.',theta_i,real(r_pe),'--',
theta_i(M), 0, 'o')
m1=min([real(t_pa) real(t_pe) real(r_pa) real(r_pe)]);
M1=max([real(t_pa) real(t_pe) real(r_pa) real(r_pe)]);
legend('t_p_a','t_p_e', 'r_p_a','r_p_e')
text(theta_i(M),0.01*(M1-m1),['\leftarrow']);
text(theta_i(M),0.018*(M1-m1),[' Brewster angle = ', num2str(theta_i(M))]);
if n1 >= n2
  text(theta_cri,-0.1*(M1-m1), [' Critial angle = ', num2str(theta_cri)]);
end
xlabel('incident angle')
axis([0 120 m1*1.1 M1*1.1])
title('Real part of coefficients')
grid on

figure(2)
plot(theta_i,imag(t_pa),'-',theta_i,imag(t_pe),':',theta_i,imag(r_pa),'-
.',theta_i,imag(r_pe),'--', theta_i(M), 0, 'o')
m1=min([imag(t_pa) imag(t_pe) imag(r_pa) imag(r_pe)]);
M1=max([imag(t_pa) imag(t_pe) imag(r_pa) imag(r_pe)]);
m11=min([abs(imag(t_pa)) abs(imag(t_pe)) abs(imag(r_pa)) abs(imag(r_pe))]);
M11=max([abs(imag(t_pa)) abs(imag(t_pe)) abs(imag(r_pa)) abs(imag(r_pe))]);
legend('t_p_a','t_p_e', 'r_p_a','r_p_e')
text(theta_i(M),0.01*(M11-m11+0.2),['\leftarrow'])
text(theta_i(M),0.03*(M11-m11+0.2),[' Brewster angle = ', num2str(theta_i(M))]);
if n1 >= n2
  text(theta_cri, -0.1*(M1-m1), [' Critial angle = ', num2str(theta_cri)]);
```

```
end
xlabel('incident angle')
axis([0 120 m1*1.1-0.1 M1*1.1+0.1])
title('Imaginary part of coefficients')
grid on

figure(3)
plot(theta_i,abs(t_pa),'-',theta_i,abs(t_pe),':',theta_i,abs(r_pa),'-.',theta_i,abs(r_pe), '--
',theta_i(M), 0, 'o')
m1=min([abs(t_pa) abs(t_pe) abs(r_pa) abs(r_pe)]);
M1=max([abs(t_pa) abs(t_pe) abs(r_pa) abs(r_pe)]);
legend('t_p_a','t_p_e', 'r_p_a','r_p_e')
text(theta_i(M),0.01*(M1-m1),['\leftarrow'])
text(theta_i(M),0.03*(M1-m1),[' Brewster angle = ', num2str(theta_i(M))]);
if n1 >= n2
  text(theta_cri, 0.1*(M1-m1), [' Critial angle = ', num2str(theta_cri)]);
end
xlabel('incident angle')
axis([0 120 m1*1.1 M1*1.1])
title('Magnitude of coefficients')
grid on

figure(4)
plot(theta_i,real(j*180/pi*log(t_pa)),'-
',theta_i,real(j*180/pi*log(t_pe)),':',theta_i,real(j*180/pi*log(r_pa)),'-
.',theta_i,real(j*180/pi*log(r_pe)), '--',theta_i(M), 0, 'o')
m1=real(min([ -j*180/pi*log(r_pa) -j*180/pi*log(r_pe)]));
M1=real(max([ -j*180/pi*log(r_pa) -j*180/pi*log(r_pe)]));
legend('t_p_a','t_p_e', 'r_p_a','r_p_e')
text(theta_i(M),0.015*(M1-m1),['\leftarrow'])
text(theta_i(M),0.03*(M1-m1),[' Brewster angle = ', num2str(theta_i(M))]);
if n1 >= n2
  text(theta_cri, -0.1*(M1-m1), [' Critial angle = ', num2str(theta_cri)]);
end
xlabel('incident angle')
axis([0 120 -185 185])
title('Phase angle of coefficients')
grid on
```

2.3 Wave Optics

In this section, we derive the *spatial frequency transfer function* for wave propagation. We then derive the important *Fresnel diffraction formula* and the *Fraunhofer diffraction formula* commonly used in Fourier optics. We start from the wave equation, Eq. (2.2-13), expressed in Cartesian coordinates:

$$\frac{1}{u^2}\frac{\partial^2 \psi}{\partial t^2} = \frac{\partial^2 \psi}{\partial x^2} + \frac{\partial^2 \psi}{\partial y^2} + \frac{\partial^2 \psi}{\partial z^2}. \qquad (2.3\text{-}1)$$

We now assume that the wavefunction $\psi(x,y,z,t)$ comprises a *complex amplitude* $\psi_p(x,y,z)$ riding on a carrier of frequency ω_0 (ψ_p is a phasor in electrical engineering):

$$\psi(x,y,z,t) = \psi_p(x,y,z)\exp(j\omega_0 t). \qquad (2.3\text{-}2)$$

Substituting Eq. (2.3-2) into Eq. (2.3-1), we get the *Helmholtz equation* for ψ_p,

$$\frac{\partial^2 \psi_p}{\partial x^2} + \frac{\partial^2 \psi_p}{\partial y^2} + \frac{\partial^2 \psi_p}{\partial z^2} + k_0^2 \psi_p = 0, \quad k_0 = \frac{\omega_0}{u}. \qquad (2.3\text{-}3)$$

At this point, we introduce 2-D Fourier transform in the next section and will thereafter use Fourier transform to find the solution to Eq.(2.3-3) for a given initial condition.

2.3.1 *Fourier transform and convolution*

The two-dimensional spatial *Fourier transform* of a square-integrable function $f(x,y)$ is given as [Banerjee and Poon (1991)]

$$F(k_x,k_y) = \int_{-\infty}^{\infty}\int_{-\infty}^{\infty} f(x,y)\exp(jk_x x + jk_y y)dxdy$$
$$= \mathcal{F}_{xy}\{f(x,y)\}. \qquad (2.3\text{-}4a)$$

$f(x, y)$ is square-integrable if

$$\int_{-\infty}^{\infty}\int_{-\infty}^{\infty} |f(x,y)|^2 \, dxdy \leq \infty.$$

The inverse Fourier transform is

$$f(x,y) = \frac{1}{4\pi^2} \int_{-\infty}^{\infty}\int_{-\infty}^{\infty} F(k_x, k_y) \exp(-jk_x x - jk_y y) dk_x dk_y$$

$$= \mathcal{F}_{xy}^{-1}\{F(k_x, k_y)\}. \tag{2.3-4b}$$

$F(k_x, k_y)$ is commonly called the *spectrum* of $f(x, y)$. The definitions for the forward and inverse transforms stated above are consistent with the engineering convention for a traveling wave, as explained in Banerjee and Poon (1991). In many optics applications, the function $f(x, y)$ represents the transverse profile of an electromagnetic or optical field at a plane z. Hence in Eqs. (2.3-4a) and (2.3-4b), $f(x, y)$ and $F(k_x, k_y)$ have z as a parameter. For instance, Eq. (2.3-4b) becomes

$$f(x,y;z) = \frac{1}{4\pi^2} \int_{-\infty}^{\infty}\int_{-\infty}^{\infty} F(k_x, k_y;z) \exp(-jk_x x - jk_y y) dk_x dk_y.$$

The usefulness of this transform lies in the fact that when substituted into the wave equation, one can reduce a three-dimensional partial differential equation (PDE) to a one-dimensional ordinary differential equation (ODE) for the spectral amplitude $F(k_x, k_y;z)$. Typical properties and examples of two-dimensional Fourier transform appear in Table 2.2.

Table 2.2 Properties and examples of some two-dimensional Fourier Transforms.

_____

Function in (x, y)

Fourier transform in (k_x, k_y)

1. $f(x, y)$

$F(k_x, k_y)$

2. $f(x - x_0, y - y_0)$

$F(k_x, k_y) \exp(jk_x x_0 + jk_y y_0)$

3. $f(ax, by)$; a, b complex constants

$\frac{1}{|ab|} F(\frac{k_x}{a}, \frac{k_y}{b})$

4. $f^*(x, y)$

$F^*(-k_x, -k_y)$

5. $F(x, y)$

$4\pi^2 f(-k_x, -k_y)$

6. $\partial f(x, y)/\partial x$

$-jk_x F(k_x, k_y)$

7. *delta function*

$\delta(x, y) = \frac{1}{4\pi^2} \int_{-\infty}^{\infty} \int_{-\infty}^{\infty} e^{\pm jk_x x \pm jk_y y} dk_x dk_y$ 1

8. 1

$4\pi^2 \delta(k_x, k_y)$

9. *rectangle function*

sinc function

$\text{rect}(x, y) = \text{rect}(x)\text{rect}(y)$,

$\text{sinc}(\frac{k_x}{2\pi}, \frac{k_y}{2\pi}) = \text{sinc}(\frac{k_x}{2\pi})\text{sinc}(\frac{k_y}{2\pi})$,

where $\text{rect}(x) = \begin{pmatrix} 1, |x| < 1/2 \\ 0, \text{otherwise} \end{pmatrix}$

where $\text{sinc}(x) = \frac{\sin(\pi x)}{\pi x}$

10. *Gaussian*

Gaussian

$\exp[-\alpha(x^2 + y^2)]$

$\frac{\pi}{\alpha} \exp[-\frac{k_x^2 + k_y^2}{4\alpha}]$

_____

==

Shifting and Scaling Properties of the Fourier Transform Example

By definition,

$$\mathcal{F}_{xy}\left\{ f(x - x_0, y - y_0) \right\} = \int_{-\infty}^{\infty} \int_{-\infty}^{\infty} f(x - x_0, y - y_0) \exp\left(jk_x x + jk_y y\right) \, dxdy \ .$$

Making change of variables $x' = x - x_0$ and $y' = y - y_0$, we have

$$\mathcal{F}_{xy}\left\{f\left(x-x_0,y-y_0\right)\right\}$$

$$= \int_{-\infty}^{\infty}\int_{-\infty}^{\infty} f(x',y')\exp\left[jk_x\left(x'+x_0\right)+jk_y\left(y'+y_0\right)\right]dx'dy'$$

$$= \exp\left(jk_x x_0 + jk_y y_0\right)\int_{-\infty}^{\infty}\int_{-\infty}^{\infty} f(x',y')\exp\left(jk_x x'+ jk_y y'\right)dx'dy'$$

$$= \exp\left(jk_x x_0 + jk_y y_0\right)F\left(k_x,k_y\right) .$$

The result above verifies item#2 in Table 2.2. Shifting the function $f(x,y)$ by x_0 and y_0 on the x-y plane adds a phase term, $\exp(jk_x x_0 + jk_y y_0)$, to its original spectrum.

For positive constants a and b, again by definition, we have

$$\mathcal{F}_{xy}\left\{f\left(ax,by\right)\right\} = \int_{-\infty}^{\infty}\int_{-\infty}^{\infty} f(ax,by)\exp\left(jk_x x + jk_y y\right)dxdy .$$

Making change of variables $x' = ax$ and $y' = by$, the above integral becomes

$$\mathcal{F}_{xy}\left\{f\left(ax,by\right)\right\} = \int_{-\infty}^{\infty}\int_{-\infty}^{\infty} f(ax,by)\exp\left(jk_x x + jk_y y\right)dxdy$$

$$= \int_{-\infty}^{\infty}\int_{-\infty}^{\infty} f(x',y')\exp[j(k_x/a)x'+ j(k_y/b)y']\frac{1}{ab}dx'dy'$$

$$= (1/ab)F(k_x/a,k_y/b) .$$

This result also holds for both a and b being negative. Similarly, we can show that for $a<0$ and $b>0$ or $a>0$ and $b<0$,

$$\mathcal{F}_{xy}\left\{f\left(ax,by\right)\right\} = -\int_{-\infty}^{\infty}\int_{-\infty}^{\infty} f(x',y')\exp[j(k_x/a)x'+ j(k_y/b)y'](dx'dy'/ab)$$

$$= (-1/ab)F(k_x/a,k_y/b).$$

Hence in general, we have

$$\mathcal{F}_{xy}\left\{f\left(ax,by\right)\right\} = (1/|ab|)F(k_x/a,k_y/b),$$

which verifies item#3 in Table 2.2. In general, a and b can be complex constants. For positive constants a and b, the function $f(ax,by)$ represents a scaled version of the original function $f(x,y)$ by factor $1/a$ along x and $1/b$ along y. Similarly, $F(k_x/a,k_y/b)$ represents the original spectrum $F(k_x,k_y)$ scaled in the spatial frequency domain by the same factors a and b.

==

==

Fourier Transform of $rect(x/x_0,y/y_0)$ Example

The one-dimensional rectangular function of $rect(x/x_0)$ is shown in the below fig. (a), where x_0 is the width of the function. The three-dimensional plot and the gray scale plot are shown in the below figs. (b) and (c), respectively. In the gray scale plot, we have assumed that an amplitude of 1 translates to "white" and an amplitude of zero to "black." The white area is $x_0 \times y_0$.

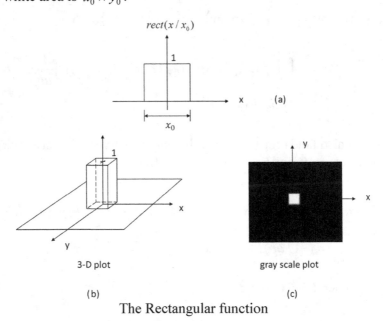

The Rectangular function

By definition,

$$\mathcal{F}_{xy}\left\{rect\left(x/x_0, y/y_0\right)\right\}$$

$$= \int_{-\infty}^{\infty}\int_{-\infty}^{\infty} rect\left(x/x_0, y/y_0\right)\exp\left(jk_x x + jk_y y\right)dxdy$$

$$= \int_{-x_0/2}^{x_0/2} 1\exp\left(jk_x x\right)dx \times \int_{-y_0/2}^{y_0/2} 1\exp\left(jk_y y\right)dy$$

$$= x_0\text{sinc}\left(\frac{x_0 k_x}{2\pi}\right)y_0\text{sinc}\left(\frac{y_0 k_y}{2\pi}\right) = x_0 y_0 \text{sinc}\left(\frac{x_0 k_x}{2\pi}, \frac{y_0 k_y}{2\pi}\right),$$

where $\text{sinc}(x) = \sin(\pi x)/\pi x$ is defined as the *sinc function*. Note that for a rectangular function of width x_0, the first zero of its spectrum is at $2\pi/x_0$. The situation is shown in the figure below. The top figures are 2-D gray-scale plots, and the bottom figures are line traces along the horizontal x-axis through the center of the top figures.

$$rect(x/x_0, y/y_0) \qquad\qquad x_0 y_0\,\text{sinc}(\frac{x_0 k_x}{2\pi}, \frac{y_0 k_y}{2\pi})$$

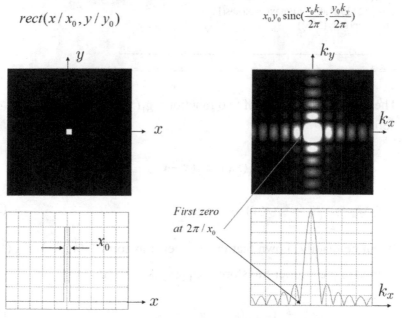

Below is the m-file generating these figures.

```
----------------------------------------------------------------
%ff2Drect.m  % Simulation of Fourier transformation of a 2D Rect function

clear

A=zeros(512,512);
B=ones(20,20);

A(256-9:256+10,256-9:256+10)=B;

Fa=fft2(A);
Fa=fftshift(Fa);

figure(1)
image(256*A)
colormap(gray(256))
axis square
axis off

figure(2)
image(2*256*abs(Fa)/max(max(abs(Fa))))
colormap(gray(256))

----------------------------------------------------------------
================================================================
```

The *convolution* $g(x,y)$ of two functions $g_1(x,y)$ and $g_2(x,y)$ is defined as

$$g(x,y) = \int_{-\infty}^{\infty} \int_{-\infty}^{\infty} g_1(x',y')g_2(x-x',y-y')dx'dy'$$

$$= g_1(x,y) * g_2(x,y). \tag{2.3-5}$$

It can be readily shown that the Fourier transform $G(k_x,k_y)$ of $g(x,y)$ is related to the Fourier transforms $G_{1,2}(k_x,k_y)$ of $g_{1,2}(x,y)$ as

$$G(k_x,k_y) = G_1(k_x,k_y)G_2(k_x,k_y). \tag{2.3-6}$$

2.3.2 *Spatial frequency transfer function and spatial impulse response of propagation*

By taking the 2-D Fourier transform, i.e., $\mathcal{F}_{xy}$, of Eq. (2.3-3) and upon some manipulations, we have

$$\frac{d^2\Psi_p}{dz^2} + k_0^2 \left(1 - \frac{k_x^2}{k_0^2} - \frac{k_y^2}{k_0^2}\right)\Psi_p = 0,$$

where $\Psi_p(k_x, k_y; z)$ is the Fourier transform of $\psi_p(x, y, z)$. We can readily solve the above equation to get

$$\Psi_p(k_x, k_y; z) = \Psi_{p0}(k_x, k_y)\exp\left[-jk_0\sqrt{1 - k_x^2/k_0^2 - k_y^2/k_0^2}\, z\right], \quad (2.3\text{-}7)$$

where

$$\Psi_{p0}(k_x, k_y) = \Psi_p(k_x, k_y; z = 0)$$

$$= \mathcal{F}_{xy}\{\psi_p(x, y, z = 0)\} = \mathcal{F}_{xy}\{\psi_{p0}(x, y)\}.$$

We can interpret Eq. (2.3-7) in the following way: Consider a linear system with $\Psi_{p0}(k_x, k_y)$ as its input spectrum (i.e., at $z = 0$) and where the output spectrum is $\Psi_p(k_x, k_y; z)$. Then, the spatial frequency response of the system is given by

$$\frac{\Psi_p(k_x, k_y; z)}{\Psi_{p0}(k_x, k_y)} = \mathcal{H}(k_x, k_y; z)$$

$$= \exp\left[-jk_0\sqrt{1 - k_x^2/k_0^2 - k_y^2/k_0^2}\, z\right]. \quad (2.3\text{-}8)$$

We will call $\mathcal{H}(k_x, k_y; z)$ the *spatial frequency transfer function of propagation* of light through a distance z in the medium.

To find the field distribution at z in the spatial domain, we take the inverse Fourier transform of Eq. (2.3-7):

$$\psi_p(x,y,z) = \mathcal{F}_{xy}^{-1}\{\Psi_p(k_x,k_y;z)\}$$

$$= \frac{1}{4\pi^2}\int\int \Psi_{p0}(k_x,k_y)\exp\left[-jk_0\sqrt{1-k_x^2/k_0^2-k_y^2/k_0^2}\,z\right]$$

$$\times\exp[-jk_xx-jk_yy]dk_xdk_y. \tag{2.3-9}$$

Now, by substituting $\Psi_{p0}(k_x,k_y) = \mathcal{F}_{xy}\{\psi_{p0}(x,y)\}$ into Eq.(2.3-9), we can express $\psi_p(x,y,z)$ as

$$\psi_p(x,y,z) = \int\int \psi_{p0}(x',y')G(x-x',y-y';z)dx'dy'$$

$$= \psi_{p0}(x,y)*G(x,y;z), \tag{2.3-10}$$

where

$$G(x,y;z) = \frac{1}{4\pi^2}\int\int \exp\left[-jk_0\sqrt{1-k_x^2/k_0^2-k_y^2/k_0^2}\,z\right]$$

$$\times\exp[-jk_xx-jk_yy]dk_xdk_y.$$

The result of Eq.(2.3-10) indicates that $G(x,y;z)$ is the *spatial impulse response of propagation* of the system. By changing of variables: $x = r\cos\theta, y = r\sin\theta, k_x = \rho\cos\phi$, and $k_y = \rho\sin\phi, G(x,y;z)$ can be evaluated as [Stark (1982)]

$$G(x,y;z) = G(r\cos\theta,r\sin\theta;z) = \tilde{G}(r;z)$$

$$= \frac{jk_0\exp\left(-jk_0\sqrt{r^2+z^2}\right)}{2\pi\sqrt{r^2+z^2}}\frac{z}{\sqrt{r^2+z^2}}\left(1+\frac{1}{jk_0\sqrt{r^2+z^2}}\right). \tag{2.3-11}$$

We can make the following observations:
(1) For $z \gg \lambda_0 = 2\pi/k_0$, i.e., we observe the field distribution many wavelengths away from the originating field, $\psi_{p0}(x,y)$, we have

$$\left(1+\frac{1}{jk_0\sqrt{r^2+z^2}}\right) \approx 1.$$

(2) $\dfrac{z}{\sqrt{r^2+z^2}} = \cos\Phi,$

where $\cos\Phi$ is called the *obliquity factor* and Φ is the angle between the positive z-axis and the line passing through the origin of the coordinates. Now, using the binomial expansion, the factor

$\sqrt{r^2+z^2} = \sqrt{x^2+y^2+z^2} \approx z + \dfrac{x^2+y^2}{2z}$, provided $x^2+y^2 \gg z^2$. This

condition is called the *paraxial approximation*, which leads to $\cos\Phi \approx 1$. If the condition is used in the more sensitive phase term and only used the first expansion term in the less sensitive denominators of the first and second terms of Eq. (2.3-11), $\tilde{G}(r;z)$ becomes the so-called *spatial impulse response*, $h(x,y;z)$, in *Fourier Optics* [Banerjee and Poon (1991), Goodman (1996), Poon (2007)]:

$$h(x,y;z) = \exp(-jk_0 z)\frac{jk_0}{2\pi z}\exp\left[\frac{-jk_0\left(x^2+y^2\right)}{2z}\right]. \qquad (2.3\text{-}12)$$

By taking the 2-D Fourier transform of $h(x,y;z)$, we have

$$H(k_x,k_y;z) = \mathcal{F}_{xy}\{h(x,y;z)\}$$

$$= \exp(-jk_0 z)\exp\left[\frac{j(k_x^2+k_y^2)z}{2k_0}\right]. \qquad (2.3\text{-}13)$$

$H(k_x,k_y;z)$ is called the *spatial frequency transfer function* in Fourier Optics [Poon 2007]. Indeed, we can derive Eq. (2.3-13) directly if we assume that $k_x^2+k_y^2 \ll k_0^2$, meaning that the x and y components of the propagation vector of a wave are relatively small. From Eq. (2.3-8), we have

$$\frac{\Psi_p(k_x,k_y;z)}{\Psi_{p0}(k_x,k_y)} = \mathcal{H}(k_x,k_y;z)$$

$$= \exp\left[-jk_0\sqrt{1-(k_x^2+k_y^2)/k_0^2}\,z\right]$$

$$\cong \exp(-jk_0 z)\exp\left[\frac{j(k_x^2+k_y^2)z}{2k_0}\right]$$

$$= H(k_x,k_y;z). \qquad (2.3\text{-}14)$$

If Eq. (2.3-12) is now used in Eq. (2.3-10), we obtain

$$\psi_p(x,y,z) = \psi_{p0}(x,y) * h(x,y;z)$$

$$= \exp(-jk_0 z)\frac{jk_0}{2\pi z}\int\int \psi_{p0}(x',y')$$

$$\times \exp\left[\frac{-jk_0}{2z}\left[(x-x')^2 + (y-y')^2\right]\right]dx'dy'. \qquad (2.3\text{-}15)$$

Equation (2.3-15) is called the *Fresnel diffraction formula* and describes the Fresnel diffraction of a beam during propagation and having an arbitrary initial complex profile $\psi_{p0}(x,y)$. The input and output planes have primed and unprimed coordinate systems, respectively. The below figure shows a block-diagram relating the input and output planes. To obtain the output field distribution $\psi_p(x,y;z)$ at a distance z away from the input field, we need to convolve the input field distribution $\psi_{p0}(x,y)$ with the spatial impulse response $h(x,y;z)$.

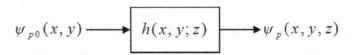

Fig. 2.5 Block diagram of wave propagation in Fourier optics.

2.3.3 *Examples of Fresnel diffraction*

Example 1: Point source

A point source is represented by $\psi_{p0}(x, y) = \delta(x)\delta(y)$. By Eq. (2.3-15), the complex field at a distance z away is given by

$$\psi_p(x, y, z) = [\delta(x)\delta(y)] * h(x, y; z)$$

$$= \frac{jk_0}{2\pi z} \exp\left[-jk_0 z - \frac{jk_0(x^2 + y^2)}{2z}\right]. \qquad (2.3\text{-}16)$$

This expression is the paraxial approximation to a *diverging spherical wave*. Now by considering the argument of the exponent in Eq. (2.3-16), we see that using the binomial expansion previously used, we can write

$$\psi_p(x, y, z) \cong \frac{jk_0}{2\pi z} \exp\left(-jk_0[z^2 + x^2 + y^2]^{\frac{1}{2}}\right)$$

$$\cong \frac{jk_0}{2\pi R} \exp(-jk_0 R), \qquad (2.3\text{-}17)$$

which corresponds to Eq. (2.2-24) for a diverging spherical wave.

Example 2: Plane Wave

For a plane wave, we write $\psi_{p0}(x, y) = 1$. Then

$$\Psi_{p0}(k_x, k_y) = 4\pi^2 \delta(k_x)\delta(k_y).$$

Using Eq. (2.3-13), we have

$$\Psi_p(k_x, k_y; z) = 4\pi^2 \delta(k_x)\delta(k_y)\exp(-jk_0 z)\exp\left[\frac{j(k_x^2 + k_y^2)z}{2k_0}\right]$$

$$= 4\pi^2 \delta(k_x)\delta(k_y)\exp(-jk_0 z).$$

Hence,

$$\psi_p(x, y, z) = \exp(-jk_0 z).$$

As the plane wave travels, it only acquires phase shift and is undiffracted, as expected.

2.3.4 *Fraunhofer diffraction*

When we examine the Fresnel diffraction pattern, which is calculated through the Fresnel diffraction formula Eq. (2.3-15), the range of applicability of this formula is from distances not too close to the source, typically from about 10 times the wavelength in practice. In this Section we examine a method of calculating the diffraction pattern at distances far away from the source or aperture. More precisely, we observe in the *far field*, that is,

$$\frac{k_0(x'^2 + y'^2)_{max}}{2} = z_R \ll z, \qquad (2.3\text{-}18)$$

where z_R is the *Rayleigh range*, then the value of the exponent $\exp[-jk_0(x'^2 + y'^2)]_{max} / 2z$ is approximately unity over the input plane (x', y'). Under this assumption, which is called the *Fraunhofer approximation*, Eq. (2.3-15) becomes

$$\psi_p(x, y, z) = \exp(-jk_0 z)\frac{jk_0}{2\pi z}\exp\left[\frac{-jk_0}{2z}(x^2 + y^2)\right]$$

$$\times \iint \psi_{p0}(x', y')\exp\left[\frac{jk_0}{z}(xx' + yy')\right]dx'dy'$$

$$= \exp(-jk_0 z)\frac{jk_0}{2\pi z}\exp\left[\frac{-jk_0}{2z}(x^2 + y^2)\right]$$

$$\times \mathcal{F}_{xy}\{\psi_{p0}(x, y)\}\Big|_{kx=k_0 x/z, ky=k_0 y/z}. \qquad (2.3\text{-}19)$$

Equation (2.3-19) is the *Fraunhofer diffraction formula* and is the limiting case of Fresnel diffraction studied earlier. The first exponential in Eq. (2.3-19) is the result of the phase change due to propagation, whereas the second exponential indicates a phase curvature that is quadratic in nature.

Note that if we are treating diffraction of red light ($\lambda_0 = 0.6328$ μm) and the maximum dimension on the input plane is $1mm$, the distance to observe the far field is then, according to Eq. (2.3-18), $z \gg 5m$.

Example: Fraunhofer diffraction of a slit of finite width

The complex amplitude of a slit illuminated by a plane wave of unity amplitude is represented by $\psi_{p0}(x,y) = rect(x/l_x)$ at the exit of the slit of width l_x along the x-direction. Note that because we are usually interested in diffracted intensities (i.e., $|\psi_p|^2$, the exponentials in Eq. (2.3-19) drop out. Furthermore, the other term beside the Fourier transform, namely, $(k_0/2\pi z)$, simply acts as a weighting factor. The intensity profile depends on the Fourier transform, and we will therefore concentrate only on this unless otherwise stated. Using

$$\mathcal{F}_{xy}\left\{rect\left(\frac{x}{l_x}\right)\right\} = l_x \text{sinc}\left(\frac{l_x k_x}{2\pi}\right) 2\pi\delta(k_y),$$

and from Eq. (2.3-19), we have

$$\psi_p(x,y;z) \propto l_x \text{sinc}\left(\frac{l_x k_0 x}{2\pi z}\right)\delta\left(\frac{k_0 y}{z}\right). \qquad (2.3\text{-}20)$$

Since there is no variation along y, we plot the normalized intensity

$$I(x)/I(0) = \text{sinc}^2\left(\frac{l_x k_0 x}{2\pi z}\right) \text{ along } x \text{ only, as shown in Fig. 2.6.}$$

Table 2.3 shows the m-file for plotting the normalized intensity.

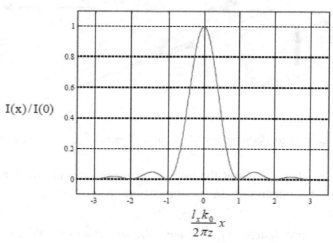

Fig. 2.6 Fraunhofer diffraction pattern of a slit.

Table 2.3 P_sinc.m: m-file for plotting $I(x)/I(0)$ for $l_x k_0 / 2\pi z = 1$.

```
%P_sinc.m Plotting of sinc^2(x) function
x= -3.5:0.01:3.5;
Sinc=sin(pi*x)./(pi*x);
plot(x,Sinc.*conj(Sinc))
axis([-3.5 3.5 -0.1 1.1])
grid on
```

We observe that the first zero of the sinc-function occurs at $x = \pm 2\pi z / l_x k_0 = \pm \lambda_0 z / l_x$, and that the angle of spread is $\theta_{spread} \cong 2\lambda_0 / l_x$ during diffraction.

In fact, we can simply estimate the spread angle from a quantum mechanical point of view [Poon and Motamedi, 1987]. Consider light emanating from an aperture of width l_x, as shown in Fig. 2.7.

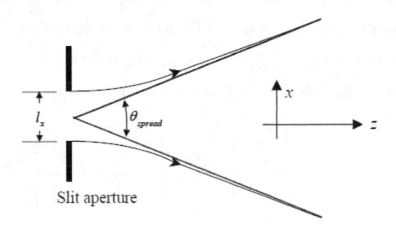

Fig. 2.7 Geometry for determination of the angle of spread θ_{spread} during diffraction.

Quantum mechanics relates the minimum uncertainty in position Δx of a quantum to the uncertainty in its momentum Δp_x according to

$$\Delta x \Delta p_x \sim \hbar. \qquad (2.3\text{-}21)$$

Now, in our problem $\Delta x = l_x$, because the *quantum* of light can emerge from any point on the aperture. Hence, by Eq. (2.3-21),

$$\Delta p_x \sim \frac{\hbar}{l_x}.$$

We define the angle of spread θ_{spread}, assumed small, as

$$\theta_{spread} \sim \frac{\Delta p_x}{p_z} \sim \frac{\Delta p_x}{p_0},$$

where p_z and p_0 represent z component of the momentum, and the momentum of the quantum, respectively. But $p_0 = \hbar k_0$, where k_0 is the propagation constant; hence,

$$\theta_{spread} \sim \frac{1}{k_0 l_x} = \frac{1}{2\pi} \frac{\lambda_0}{l_x}, \tag{2.3-22}$$

where λ_0 denotes the wavelength in the medium of propagation of the light. Thus, the angle of spread is inversely proportional to the aperture width, l_x.

2.3.5 *Fourier transforming property of an ideal lens*

Since a lens is a phase object, and for an ideal focusing lens of focal length f, its phase transformation function, $t_f(x, y)$, is given by

$$t_f(x, y) = \exp\left[j\frac{k_0}{2f}(x^2 + y^2) \right]. \tag{2.3-23}$$

The reason for this is that for a uniform plane wave incident upon the lens, the wavefront behind the lens is a converging spherical wave (for $f > 0$) that converges ideally to a point source a distance $z = f$ behind the lens. Upon comparing with the paraxial approximation for a diverging spherical wave, as given by Eq.(2.3-16), Eq. (2.3-23) readily follows for an ideal thin lens (thickness of the lens being zero and the x and y dimensions being infinite).

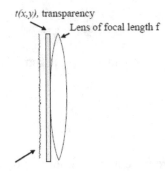

t(x,y), transparency

Lens of focal length f

Incident Complex Field

Fig. 2.8 A transparency immediately before an ideal lens under complex field illumination.

Let us now investigate the effect of placing a transparency $t(x, y)$ against the ideal lens, as shown in Fig. 2.8. In general, $t(x, y)$ is a complex function such that if a complex field $\psi_p(x, y)$ is incident on it, the field immediately behind the transparency-lens combination is given by

$$\psi_p(x, y)t(x, y)t_f(x, y) = \psi_p(x, y)t(x, y)\exp\left[j\frac{k_0}{2f}(x^2 + y^2)\right],$$

where we have assumed that the transparency is infinitely thin, as is the case for the ideal lens. Then, for brevity, under illuminated by a unit-amplitude plane wave, i.e., $\psi_p(x, y) = 1$, the field immediately behind the combination is given by $t(x, y)\exp\left[j\frac{k_0}{2f}(x^2 + y^2)\right]$. We then find the field distribution at a distance $z = f$ by using the Fresnel diffraction formula, Eq. (2.3-15), as

$$\psi_p(x, y, z = f) = \exp(-jk_0 f)\frac{jk_0}{2\pi f}\exp\left[\frac{-jk_0}{2f}(x^2 + y^2)\right]$$

$$\times \int\int t(x', y')\exp\left[j\frac{k_0}{f}(xx' + yy')\right]dx'dy'$$

$$= \exp(-jk_0 f)\frac{jk_0}{2\pi f}\exp\left[\frac{-jk_0}{2f}(x^2 + y^2)\right]$$

$$\times \mathcal{F}_{xy}\{t(x, y)\}\Big|_{kx=k_0 x/f, ky=k_0 y/f}, \tag{2.3-24}$$

where x and y denote the transverse coordinates at $z = f$. Hence, the complex field on the focal plane $(z = f)$ is proportional to the Fourier transform of $t(x, y)$ but with a phase curvature. Note that if $t(x, y) = 1$, i.e., the transparency is totally clear, we have $\psi_p(x, y, z = f) \propto \delta(x, y)$, which corresponds to the focusing of a plane wave by a lens. Note that, for an ideal divergent lens, its phase transformation function is given by

$$\exp\left[-j\frac{k_0}{2f}(x^2 + y^2)\right].$$

All physical lenses have finite apertures and we can model this physical situation as a lens with an infinite aperture followed immediately by a transparency described by what is called the *pupil function* $p_f(x, y)$ of the lens. Typical pupil functions are $rect(x / X, y / Y)$ or $circ\ (r / r_0)$, where X, Y, and r_0 are some constants, and $r = (x^2 + y^2)^{1/2}$ and that $circ\ (r / r_0)$ denotes a value 1 within a circle of radius r_0 and 0 otherwise. Hence, if we have transparency $t(x, y)$ against a lens with a finite aperture, the field at the back focal plane of the lens is given by

$$\psi_p(x, y, z = f) \propto \mathcal{F}_{xy}\left\{t(x, y)p_f(x, y)\right\}\big|_{kx = k_0 x / f, ky = k_0 y / f} \qquad (2.3\text{-}25)$$

under plane wave illumination.

Example: Transparency in front of a lens

Suppose that a transparency $t(x, y)$ is located at a distance d_0 in front of a convex lens with an infinitely large aperture and is illuminated by a plane wave of unit strength as shown in Fig. 2.9. The physical situation is shown in Fig. 2.9(a), which can be represented by a block diagram given by Fig. 2.9(b). According to the block diagram, we write

$$\psi_p(x, y; f) = \{[t(x, y) * h(x, y; d_0)]t_f(x, y)\} * h(x, y; f), \qquad (2.3\text{-}26)$$

which can be evaluated to become

$$\psi_p(x,y;f) = \frac{jk_0}{2\pi f} \exp[-jk_0(d_0 + f)] \exp\left[-j\frac{k_0}{2f}\left(1 - \frac{d_0}{f}\right)(x^2 + y^2)\right]$$

$$\times \mathcal{F}_{xy}\{t(x,y)\}\Big|_{kx=k_0x/f, \, ky=k_0y/f} \, . \qquad (2.3\text{-}27)$$

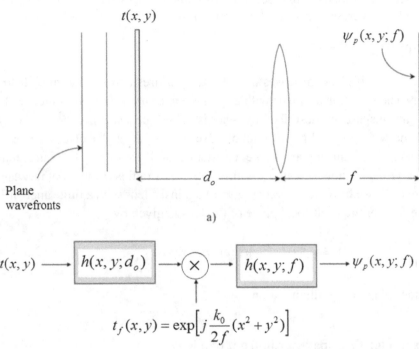

Fig. 2.9 Plane-wave illumination of a transparency $t(x, y)$ located distance d_0 in front of a converging lens of focal length f: (a) Physical situation and (b) Block diagram.

    Note that, as in Eq.(2.3-24), a phase curvature factor again precedes the Fourier transform, but vanishes for the special case $d_0 = f$. Thus, *when the transparency is placed in the front focal plane of the convex lens the phase curvature disappears, and we recover the exact Fourier transform on the back focal plane.* Fourier processing on an "input" transparency located on the front focal plane may now be performed on the back focal plane. Thus, a lens brings the Fraunhofer diffraction pattern (usually seen in the far field) to its back focal plane through the quadratic phase transformation of the ideal lens. This is the essence of Fourier optics to perform coherent image processing.

==

Coherent Image Processing Example

In the figure of the example, we show a standard two-lens coherent image processing system. The system is known as the *4-f optical system* for optical image processing. On the object plane, we have an input of the form of a transparency, $t(x,y)$, which is illuminated normally by a plane wave of unit amplitude. A transparency function, $p(x,y)$, is placed on the confocal plane of the optical system. $p(x,y)$ is called the *pupil function* of the processing system. Our objective is to find the field distribution, $\psi_{pi}(x,y)$, on the image plane.

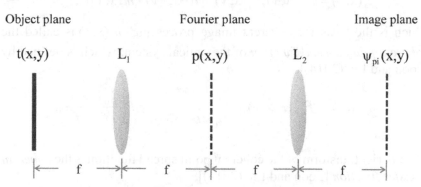

| Object plane | Fourier plane | Image plane |

t(x,y)　　　L$_1$　　　p(x,y)　　　L$_2$　　　ψ_{pi}(x,y)

$$\leftarrow \quad f \quad \rightarrow\!\!\ast\!\!\leftarrow \quad f \quad \rightarrow\!\!\ast\!\!\leftarrow \quad f \quad \rightarrow\!\!\ast\!\!\leftarrow \quad f \quad \rightarrow$$

Two-lens coherent image processing system

According to Eq. (2.3-27) by setting $d_0 = f$ and neglecting some constant, we have the spectrum of the object on the back focal plane of lens L1, given by

$$\mathcal{F}_{xy}\{t(x,y)\}\Big|_{k_x=k_0 x/f,\,k_y=k_0 y/f} = T\left(\frac{k_0 x}{f},\frac{k_0 y}{f}\right).$$

The spectrum of the input object is now modified by the pupil function as the field distribution immediately after the pupil function is

$$T\left(\frac{k_0 x}{f},\frac{k_0 y}{f}\right)p(x,y).$$

According to Eq. (2.3-27) again, this field will be Fourier transformed to give the field on the image plane as

$$\psi_{pi}(x,y) = \mathcal{F}_{xy}\left\{T\left(\frac{k_0 x}{f},\frac{k_0 y}{f}\right)p(x,y)\right\}\Bigg|_{k_x = k_0 x/f, k_y = k_0 y/f},$$

which can be evaluated, in terms of convolution, to give

$$\psi_{pi}(x,y) = t(-x,-y) * \mathcal{F}_{xy}\{p(x,y)\}\big|_{k_x = k_0 x/f, k_y = k_0 y/f}$$

$$= t(-x,-y) * h_c(x,y).$$

The corresponding image intensity is

$$I_i(x,y) = \psi_{pi}(x,y)\psi_{pi}^*(x,y) = |t(-x,-y) * h_c(x,y)|^2,$$

which is the basis for coherent image processing. $h_c(x,y)$ is called the *coherent point spread function* of the optical system, which is defined by [Poon and Liu (2014)]

$$h_c(x,y) = \mathcal{F}_{x,y}\{p(x,y)\}\big|_{k_x = k_0 x/f, k_y = k_0 y/f} = P\left(\frac{k_0 x}{f},\frac{k_0 y}{f}\right).$$

The Fourier transform of the coherent point spread function is the *coherent transfer function* [Poon and Liu (2014)],

$$H_c(k_x, k_y) = \mathcal{F}_{x,y}\{h_c(x,y)\} = p\left(\frac{-fk_x}{k_0},\frac{-fk_y}{k_0}\right).$$

Note that for all-pass filtering, i.e., when $p(x,y) = 1$, $h_c(x,y) \propto \delta(x,y)$ and $H_c(k_x, k_y) = $ constant . The field distribution on the image plane becomes

$$\psi_{pi}(x,y) = t(-x,-y) * h_c(x,y) = t(-x,-y) * \delta(x,y) = t(-x,-y).$$

The negative sign in the argument of $t(x,y)$ represents that the original input on the image plane has been flipped and inverted on the image plane, consistent with ray-optics interpretation of the situation.

===

===

Low-Pass Filtering with the 4-f System: MATLAB Example

We take $p(x,y) = circ(r / r_0)$ and according to the result from the last example, we field distribution on the image plane is

$$\psi_{pi}(x,y) = \mathcal{F}_{xy}\left\{ T\left(\frac{k_0 x}{f}, \frac{k_0 y}{f}\right) p(x,y) \right\}\bigg|_{k_x = k_0 x / f, k_y = k_0 y / f} .$$

$$= \mathcal{F}_{xy}\left\{ T\left(\frac{k_0 x}{f}, \frac{k_0 y}{f}\right) circ(r / r_o) \right\}\bigg|_{k_x = k_0 x / f, k_y = k_0 y / f}$$

For this kind of pupil chosen, filtering is of lowpass characteristic as the opening of the disk on the pupil plane only allows the low spatial frequencies to physically go though. In the below Figs. (a) and (d), we show the original of the image and its spectrum. In Figs. (b) and (c), we show the filtered images, and their corresponding lowpass filtered spectra are shown in Figs. (e) and (f), respectively, where the lowpass filtered spectra are obtained by multiplying the original spectrum by $circ(r / r_0)$. Note that the radius r_0 used in (f) is smaller than that used in Fig. (e), signifying that the system is restricting more spatial frequencies from going through the optical system and hence the filtered image on the image plane [Fig. (c)] is more blurred than that shown in Fig. (b). The MATLAB code used to generated these results is listed at the end of this example. More filtering examples and simulations can be found in Poon and Liu [2014].

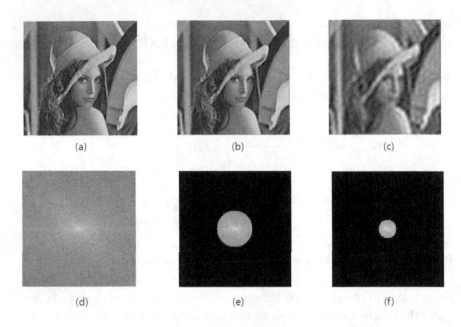

(a)   (b)   (c)

(d)   (e)   (f)

Lowpass filtering examples

```
% Low-pass filtering of an image
clear all;
A=imread('lena.jpg');
A=double(A);
SP=fftshift(fft2(A));
D=log(abs(SP));

figure (1)
image(256*A/max(max(A)));
colormap(gray(256))
title('Original image'); axis square; axis off

figure (2)
image(256*D(256-127:256+128,256-127:256+128 )/max(max(D(256-127:256+128,256-
127:256+128)))); % spectrum
colormap(gray(256))
title('Original spectrum'); axis square; axis off

a=1:512;
```

```
b=1:512;
[A, B]=meshgrid(a, b);
Cir=((A-257).^2+(B-257).^2);
filter=Cir <= 20^2;
filter=double(filter);

G=log(abs(filter.*SP));
figure(3);
image(256*G(256-127:256+128,256-127:256+128)/max(max(G(256-127:256+128,256-
127:256+128))))
colormap(gray(256))
title('Low-pass spectrum'); axis square; axis off

figure(4);
SPF=SP.*filter;
E=abs(ifft2(fftshift(SPF)));
image(256*E./max(max(abs(E))))
colormap(gray(256))
title('Low-pass image'); axis square; axis off
```

===

2.3.6 *Resonators and Gaussian beams*

In Chapter 1, we analyzed an optical resonator using geometrical optics. In this section, we will use wave optics analysis to find the beam profile of different *modes* in the resonator, and hence the field distributions of the output light beam. For simplicity, we consider a confocal resonator system (see Section 1.5) that consists of a pair of concave mirrors of equal radii, i.e., $R = d$ and $R = -d$, separated by a distance d. The resonator system is again shown in Fig. 2.10(a).

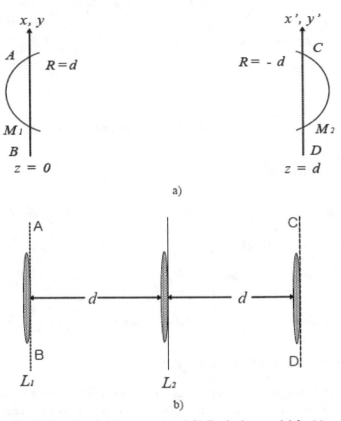

a)

b)

Fig. 2.10 (a) Confocal resonator, and (b) Equivalent model for (a).

In order to calculate the transverse modes inside the resonator, we let $\psi_p(x,y)$ represent the optical field along plane AB as shown in Fig. 2.10(a). This field $\psi_p(x,y)$ undergoes Fresnel diffraction over a distance d, is then reflected from mirror M_2, travels back to mirror M_1, and undergoes a second reflection in one round trip. $\psi_p(x,y)$ is called a mode of the resonator if it reproduces itself after one round trip (apart from some constant). From geometry, it is easy to see that AB and CD (which are separated by d) are symmetric about $z = d/2$, and hence the resonator in Fig. 2.10(a) may be "unfolded" as shown in Fig. 2.10(b) to analyze the situation. In this equivalent model, the "unfolding" implies that the field $\psi_p''(x,y)$ immediately behind lens L_2 is identical to the field $\psi_p(x,y)$ immediately behind lens L_1. The focal length f of each lens must be equal to $d/2$. Therefore, we write

$$\psi_p''(x,y) = \gamma\psi_p(x,y), \qquad (2.3\text{-}28)$$

where γ is some complex constant. Equation (2.3-28) formulates an eigenvalue problem. Explicitly, Eq.(2.3-28) reads [see Eq.(2.3-15)]:

$$\gamma\psi_p(x,y) = \left[\frac{jk_0}{2\pi d}\exp(-jk_0 d)\int\int_S \psi_p(x',y') \right.$$

$$\times\exp\left(-j\frac{k_0}{2d}\left[(x-x')^2 + (y-y')^2\right]\right)dx'dy' \Bigg]$$

$$\times\exp\left[j\frac{k_0}{d}(x^2+y^2) \right]$$

$$= \frac{jk_0\exp(-jk_0 d)}{2\pi d}\exp\left[j\frac{k_0}{2d}(x^2+y^2) \right]\int\int_S \psi_p(x',y')$$

$$\times\exp\left[-j\frac{k_0}{2d}(x'^2+y'^2 - 2xx' - 2yy') \right]dx'dy', \quad (2.3\text{-}29)$$

where S is the area of mirrors M_1 and M_2 in Fig. 2.10(a). Equation (2.3-29) is an integral equation with $\psi_p(x,y)$ to be solved.

As we will see shortly, there are an infinite number of solutions $\psi_{m,n}$ called *eigenfunction*, or *eigenmodes*, each with an associated eigenvalue γ_{mn}, where m and n denote the *transverse mode numbers*. The mode numbers determine the transverse field distribution of the mode. The eigenvalues γ_{mn} have physical meanings. If we put

$$\gamma_{mn} = |\gamma_{mn}|e^{j\phi_{mn}}, \qquad (2.3\text{-}30)$$

we see that quantity $1 - |\gamma_{mn}|^2$ gives the energy loss per half-cycle transit. The loss is called the diffraction loss of the resonator and is due to the "energy spills" around the reflecting mirrors [because we are considering the so called open-sided resonators, see Fig. 2.10(a), in which we assume that S is finite]. For mirrors with very large aperture, i.e., $S \to \infty$, the field does not lose power during round trips, and hence γ_{mn} is simple a phase factor. Now, the phase of γ_{mn}, i.e., ϕ_{mn}, is the phase shift per half-cycle transit, which determines the oscillation frequencies of the resonator.

Returning to Eq. (2.3-29) and defining a function $f(x, y)$ such that

$$f(x, y) = \psi_p(x, y) \exp\left[j\frac{k_0}{2d}(x^2 + y^2) \right],$$

Eq. (2.3-29) becomes

$$\gamma f(x, y) = \frac{jk_0 \exp(-jk_0 d)}{2\pi d} \int_{-a}^{a} \int_{-a}^{a} f(x', y') \exp\left[j\frac{k_0}{d}(xx' + yy') \right] dx' dy',$$

$$(2.3\text{-}31)$$

where we have assumed that the two mirrors have square cross-sections of linear dimension $2a$.

Introducing the dimensionless variables ζ and η through the relations

$$\zeta = \sqrt{2\pi N}\frac{x}{a} \qquad \text{and} \qquad \eta = \sqrt{2\pi N}\frac{y}{a},$$

where $N = \dfrac{a^2 k_0}{2\pi d}$ is called the *Fresnel number* of the square aperture,

Eq.(2.3-31) reduces to

$$\gamma f(\zeta, \gamma) = \frac{je^{-jk_0 d}}{2\pi} \int_{-\sqrt{2\pi N}}^{\sqrt{2\pi N}} \int_{-\sqrt{2\pi N}}^{\sqrt{2\pi N}} f(\zeta', \eta') e^{j(\zeta\zeta' + \eta\eta')} d\zeta' d\gamma'.$$

To solve for $f(\zeta, \eta)$, we use the *separation of variables technique* and write

$$\gamma = \gamma_1 \gamma_2 \quad \text{and} \quad f(\zeta, \eta) = p(\zeta)q(\eta). \qquad (2.3\text{-}32)$$

On substitution, Eq.(2.3-32) can be written as two separate equations,

$$\gamma_1 p(\zeta) = \exp\left(-\frac{jk_0 d}{2} \right) \sqrt{\frac{j}{2\pi}} \int_{-\sqrt{2\pi N}}^{\sqrt{2\pi N}} p(\zeta') \exp(j\zeta\zeta') d\zeta' \qquad (2.3\text{-}33a)$$

and

$$\gamma_2 q(\eta) = \exp\left(-\frac{jk_0d}{2}\right)\sqrt{\frac{j}{2\pi}}\int_{-\sqrt{2\pi N}}^{\sqrt{2\pi N}} q(\eta')\exp(j\eta\eta')d\eta'. \quad (2.3\text{-}33b)$$

The solutions of these integral equations are *prolate spheroidal functions* [Slepian and Pollack (1961)] and these functions are numerically tabulated. Here we will only consider the case when $N \gg 1$. For such a case, we can extend the limits of integration in Eq. (2.3-33) from $-\infty$ to ∞. Equation (2.3-33) tells us that the functions $p(\zeta)$ and $q(\eta)$ are their own Fourier transforms apart from some constant. These functions are called *self-Fourier transform functions* [Banerjee and Poon, 1995]. A complete set of functions satisfies the condition is the Hermite-Gaussian functions.

The Hermit-Gaussian function $g_m(\zeta)$ is defined as follows:

$$j^m g_m(\zeta) = \frac{1}{\sqrt{2\pi}}\int_{-\infty}^{\infty} g_m(\zeta')e^{j\zeta\zeta'}d\zeta', \quad (2.3\text{-}34)$$

where $g_m(\zeta) = H_m(\zeta)e^{-\zeta^2/2}$ and $H_m(\zeta)$ is the Hermite polynomial of order m, defined by

$$H_m(\zeta) = (-1)^m e^{\zeta^2} \frac{d^m}{d\zeta^m} e^{-\zeta^2}. \quad (2.3\text{-}35a)$$

The first five Hermite polynomials are

$$H_0(\zeta) = 1$$
$$H_1(\zeta) = 2\zeta$$
$$H_2(\zeta) = 4\zeta^2 - 2$$
$$H_3(\zeta) = 8\zeta^3 - 12\zeta$$
$$H_4(\zeta) = 16\zeta^4 - 48\zeta^2 + 12. \quad (2.3\text{-}35b)$$

Figure 2.11 shows the three lowest-order Hermite-Gaussian functions. Note that, in general, the m th-order function contains m nulls.

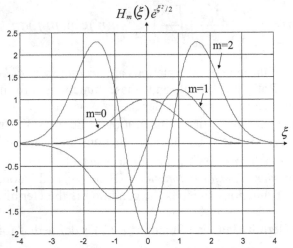

Fig. 2.11 The three lowest-order Hermite-Gaussian functions.

In order that $p(\zeta)$ and $q(\eta)$ be of the Hermite-Gaussian form, it is required that

$$\gamma_1 = j^{(m+1/2)} \exp\left(-\frac{jk_0 d}{2}\right) = \exp\left\{-j\left[\frac{1}{2}k_0 d - \frac{1}{2}\left(m+\frac{1}{2}\right)\pi\right]\right\}$$

and

$$\gamma_2 = j^{(n+1/2)} \exp\left(-\frac{jk_0 d}{2}\right) = \exp\left\{-j\left[\frac{1}{2}k_0 d - \frac{1}{2}\left(n+\frac{1}{2}\right)\pi\right]\right\}.$$

(2.3-36)

The solutions of Eq. (2.3-33) are, therefore, the Hermit-Gaussian functions

$$p_m(\zeta) = H_m(\zeta)e^{-\zeta^2/2} \qquad (2.3\text{-}37a)$$

and

$$q_n(\eta) = H_n(\eta)e^{-\eta^2/2}, \qquad (2.3\text{-}37b)$$

where m and n denote the transverse mode numbers and determine the field distribution of the mode. Thus the complete solution of Eq. (2.3-29) with $N \gg 1$ can be expressed as Hermit-Gaussian beams,

$$\psi_{Pmn}(x,y) = A\exp\left[j\frac{1}{2}(\zeta^2 + \eta^2)\right]P_m(\zeta)q_n(\eta)\Bigg|_{\zeta=\sqrt{2\pi N}x/a,\,\eta=\sqrt{2\pi N}y/a}$$

$$= A\exp\left[j\frac{k_0}{2d}(x^2 + y^2)\right]H_m\left(\sqrt{\frac{k_0}{d}}x\right)H_n\left(\sqrt{\frac{k_0}{d}}y\right)$$

$$\times \exp\left[-\frac{k_0}{2d}(x^2 + y^2)\right], \tag{2.3-38}$$

where A is some constant. Each set (m,n) corresponds to a particular transverse electromagnetic (TEM) mode of the resonator as the electric (and magnetic) field of the electromagnetic wave is orthogonal to the resonator Z-axis. The lowest-order Hermite polynomial H_0 is equal to unity; hence the mode corresponding to the set $(0,0)$ is called the TEM_{00} mode and has a Gaussian radial profile.

Figure 2.12 depicts the intensity patterns $|\psi_{Pmn}(x,y)|^2$ of some modes, and Table 2.3 provides the m-file for plotting them. Note that higher-order modes have a more spread-out intensity distribution and therefore would have higher diffraction losses. Most practical lasers are made to oscillate in the TEM_{00} mode. Because higher-order modes have wide transverse dimensions, they can be suppressed by placing a circular aperture inside the resonator.

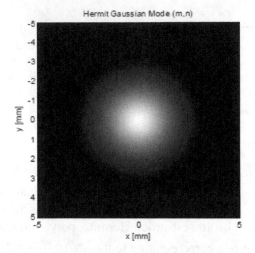

a) Mode $m = 0$, $n = 0$

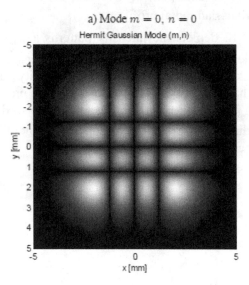

b) Mode $m = 3$, $n = 3$

Fig. 2.12 Intensity patterns for two modes.

Table 2.4 HG.m (m-file for plotting intensity patterns of Hernite-Gaussian modes).

```
%HG.m (Plotting Hermit-Gaussians up to m=n=3 mode)
clear
m=input('m (enter between 0 to 3) = ');
n=input('n (enter between 0 to 3) = ');
%waist=(ko/d)^0.5=1 [mm]
ko_d=1;
Xmin=-5;
Xmax=5;
Step_s=0.0001*300/(1.276*1.2);
x=Xmin:Step_s:Xmax;
y=x;
if m==0
  Hm=ones(size(x));
end
if m==1
  Hm=2*x*ko_d;
end
if m==2
  Hm=4*x.^2-2*ones(size(x));
end
if m==3
  Hm=8*x.^3-12*x;
end
if n==0
  Hn=ones(size(x));
end
if n==1
  Hn=2*x*ko_d;
end
if n==2
  Hn=4*x.^2-2*ones(size(x));
end
if n==3
  Hn=8*x.^3-12*x;
end
%Length of consideration range
L=Xmax-Xmin;
n=size(x);
N=n(2);
for k=1:N
  for l=1:N
    psy(k,l)=exp(j*ko_d/2*(x(l)^2+y(k)^2))*Hm(l)*Hn(k)*exp(-
ko_d/2*(x(l)^2+y(k)^2));
end


end
figure(1)
```

```
image(x,y,256*abs(psy)/max(max(abs(psy))))
colormap(gray(256))
title('Hermit Gaussian Mode (m,n)')
xlabel('x [mm]')
ylabel('y [mm]')
axis square
```

So far we have discussed only the eigenfunctions of Eqs. (2.3-33 (a) and (b)). In discussing their corresponding eigenvalues, we note that from Eq. (2.3-36),

$$\gamma = \gamma_1 \gamma_2 = \exp\left\{-j\left[k_0 d - (m+n+1)\frac{\pi}{2}\right]\right\}. \qquad (2.3-39)$$

Observe that $|\gamma| = 1$, implying that the diffraction losses are zero. This is expected, as in our analysis we have essentially assumed the mirror cross-section is of large dimensions, that is, $N \gg 1$. We have mentioned previously that the phase of γ represents the additional phase shift per half-cycle transit. The condition for supporting a mode is that the field does not change of the field in a round trip through the resonator. This means that the phase change of the field in a round trip should be an integral multiple of 2π, or for the half-cycle transit, the phase change must be an even multiple of π. Therefore, we must have

$$k_{lmn} d - (m+n+1)\frac{\pi}{2} = l\pi, \quad l = 1,2,3,\ldots, \qquad (2.3-40)$$

where l refers to the *longitudinal mode number*. Hence, the resonant frequencies of the oscillation of the resonator are expressible as

$$\omega_{lmn} = \pi(2l+m+n+1)\frac{u}{2d}, \qquad (2.3-41)$$

where we have used $k_{lmn} = \omega_{lmn} / u$ and u is the velocity in the medium Those frequencies that satisfy the above equation are allowed in the resonator. Note that modes having the same value of $2l+m+n$ have the same resonance frequency, although they have different field distributions. These modes are *degenerate*. The frequency spacing with the same value of m and n and with l differing by 1 is

$$\omega_{l+1} - \omega_l = \Delta\omega_l = \frac{\pi u}{d}, \qquad (2.3-42)$$

and the frequency separation between two transverse modes having the same value of l is

$$\Delta\omega_m = \frac{\pi u}{2d} = \Delta\omega_n. \qquad (2.3\text{-}43)$$

The field distribution given by Eq.(2.3-38) represents the field along plane AB as shown in Fig. (2.13), and the field distribution in the plane midway between the two mirrors can be evaluated, using the Fresnel integral formula as

$$\psi_{Pmn}\big|_{z=d/2} = j^{m+n+1} A \exp\left(-\frac{jk_0 d}{2}\right) \exp\left[-\frac{(x^2+y^2)}{w_0^2}\right]$$
$$\times H_m\left(\frac{\sqrt{2}x}{w_0}\right) H_n\left(\frac{\sqrt{2}y}{w_0}\right), \qquad (2.3\text{-}44)$$

where $w_0 = \sqrt{d/k_0}$ is called the *waist* of the beam. Note that at the waist the beam has no phase curvature, and therefore, the phase fronts are planar midway between the mirrors. Also from Eq.(2.3-38), we note that the phase fronts of the field distribution at mirrors M_1 and M_2 have radii of curvature that are identical to the radii of the mirrors.

In practice, the laser output comprises a small fraction of the energy in the resonator that is coupled out through mirror M_2, which is made partially refractive. We will be interested to know the behavior of the laser output with propagation. To find this, it is convenient to have the resonator center taken to be the origin $z=0$, so that the field at $z=0$ (see Fig. 2.13) can be expressed as

$$\psi_{Pmn}(x,y,z=0) = E_0 \exp\left[-\frac{(x^2+y^2)}{w_0^2}\right] H_m\left(\frac{\sqrt{2}x}{w_0}\right) H_n\left(\frac{\sqrt{2}y}{w_0}\right). \qquad (2.3\text{-}45)$$

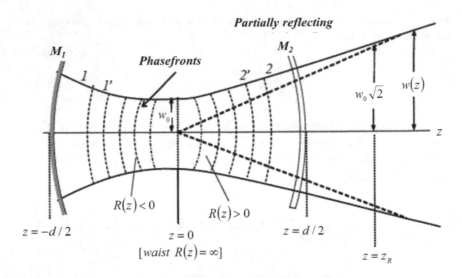

Fig. 2.13 Confocal resonator system.

One can employ the Fresnel integral formula to find the field at any plane z inside as well as outside the resonator:

$$\psi_{Pmn}(x,y,z) = \frac{E_0 w_0}{w(z)} \exp(-jk_0 z) \exp\left[-\frac{jk_0(x^2+y^2)}{2R(z)}\right]$$

$$\times \exp[-j(m+n+1)\phi(z)] H_m\left(\frac{\sqrt{2}x}{w(z)}\right) H_n\left(\frac{\sqrt{2}y}{w(z)}\right) \exp\left[-\frac{(x^2+y^2)}{w^2(z)}\right],$$

$$(2.3\text{-}46)$$

where $w(z), \phi(z),$ and $R(z)$ are defined as follows:

$$w^2(z) = w_0^2\left[1+(\frac{z}{z_R})^2\right],$$

$$R(z) = (z^2 + z_R^2)/z;$$

$$\phi(z) = -\tan^{-1}(z/z_R)$$

with $z_R = d/2$. We remark that the resonant frequencies can also be calculated through the function $\phi(z)$. In fact, by evaluating the phase term, namely, $k_0 z + (m + n + 1)\phi(z)$, at $z = \pm d / 2$, and setting the difference equal to $l\pi$, we have

$$k_{lmn} d + (m + n + 1)\left[\phi\left(\frac{d}{2}\right) - \phi\left(-\frac{d}{2}\right)\right] = l\pi, \quad k_{lmn} = \frac{\omega_{lmn}}{u},$$

which can be evaluated and shown to be identical to Eq. (2.3-40).

2.4 Gaussian Beam Optics and MATLAB Examples

In this Section, we will study the propagation or Fresnel diffraction of a Gaussian beam. We consider a Gaussian beam in two transverse dimensions with initially planar wavefronts:

$$\psi_{p0}(x, y) = \exp[-(x^2 + y^2) / w_0^2], \tag{2.4-1}$$

where w_0 is called the *waist* of the Gaussian beam. The Fourier transform of this is

$$\Psi_{p0}(k_x, k_y) = \pi w_0^2 \exp[-(k_x^2 + k_y^2) w_0^2 / 4]. \tag{2.4-2}$$

Using Eq. (2.3-14), the spectrum after propagation by a distance z is given by

$$\begin{aligned}
\Psi_p(k_x, k_y, z) &= \Psi_{p0}(k_x, k_y) \exp(-jk_0 z) \exp[j(k_x^2 + k_y^2) z / 2k_0] \\
&= \pi w_0^2 \exp[-(k_x^2 + k_y^2) w_0^2 / 4] \exp(-jk_0 z) \exp[j(k_x^2 + k_y^2) z / 2k_0] \\
&= \pi w_0^2 \exp(-jk_0 z) \exp[j(k_x^2 + k_y^2) q / 2k_0],
\end{aligned} \tag{2.4-3}$$

where q is called the q-*parameter* of the Gaussian beam, defined as

$$q = z + jz_R, \tag{2.4-4}$$

with z_R defined as the *Rayleigh range* of the Gaussian beam:

$$z_R = k_0 w_0^2 / 2. \tag{2.4-5}$$

The shape of the beam after propagation through a distance z can be found by taking the inverse Fourier transform of Eq. (2.4-3):

$$\psi_p(x,y,z) = \exp(-jk_0 z)\frac{jk_0 w_0^2}{2q}\exp[-jk_0(x^2+y^2)/2q], \quad (2.4\text{-}6a)$$

which can be written as

$$\psi_p(x,y,z) = \frac{w_0}{w(z)} e^{-(x^2+y^2)/w^2(z)}$$

$$\times e^{-jk_0(x^2+y^2)/2R(z)} e^{-j\phi(z)} e^{-jk_0 z}, \quad (2.4\text{-}6b)$$

where the functional forms of $w(z), \phi(z)$, and $R(z)$ have been defined in Eq. (2.3-46) with z_R now defined by Eq. (2.45). Note, from Eq. (2.4-6b), that

(1) the width $w(z)$ of the Gaussian beam is a monotonically increasing function of propagation z, and reaches $\sqrt{2}$ times its original width or waist w_0 at $z = z_R$, the Rayleigh range,

(2) the radius of curvature $R(z)$ of the phase fronts is initially infinite, corresponding to an initially planar wavefronts, as defined by Eq.(2.4-1), but reaches a minimum value of $2z_R$ at $z = z_R$, before starting to increase again. This makes sense, since far from the source $z = 0$, and well past the Rayleigh range, the Gaussian beam resembles a spherical wavefront, with the radius of curvature approaching z, the distance of propagation,

(3) the slowly varying phase $\phi(z)$, monotonically varied from 0 at $z = 0$ to $-\pi/2$ as $z \to \infty$, with a value of $-\pi/2$ at $z = z_R$;

(4) the angular spread of the Gaussian beam, θ_{sp}, can be obtained by $\tan\theta_{sp} = w(z)/z$, as it is clear from the geometry shown in Fig. 2.13. For large z, the angular spread becomes

$$\theta_{sp} \approx \frac{\lambda_0}{\pi w_0}, \quad (2.4\text{-}7)$$

as $w(z) \approx 2z / k_0 w_0$.

Finally, we want to point out that the q-parameter is a useful quantity to know for a Gaussian beam. We can write, according to Eq. (2.4-4),

$$\frac{1}{q} = \frac{1}{z + jz_R} = \frac{1}{R(z)} - j\frac{2}{k_0 w^2(z)}. \qquad (2.4\text{-}8)$$

Note that the q-parameter contains all the information about the Gaussian, namely its curvature $R(z)$ and its waist $w(z)$. Indeed, if we know how the q transforms, we know how the Gaussian beam behaves.

2.4.1 q-transformation of Gaussian beams

The q-parameter of a Gaussian beam makes it real convenient to track an arbitrary Gaussian beam during its propagation through an optical system. Consider, for instance, the propagation of a Gaussian beam through a distance d. From (2.4-3) it is easy to see in the spatial frequency domain, propagation by a distance d amounts to multiplying the spectrum at z by an exponential term $\exp[j(k_x^2 + k_y^2)d / 2k_0]$, and a constant factor $\exp[-jk_0 d]$,

$$\Psi_p(k_x, k_y, z + d) = \Psi_p(k_x, k_y; z)e^{j(k_x^2 + k_y^2)d/2k_0}e^{-jk_0 d}$$
$$= \pi w_0^2 e^{j(k_x^2 + k_y^2)q/2k_0}e^{j(k_x^2 + k_y^2)d/2k_0}e^{-jk_0(z+d)}$$
$$= \pi w_0^2 e^{j(k_x^2 + k_y^2)q_d/2k_0}e^{-jk_0(z+d)}. $$

$$(2.4\text{-}9)$$

Thus, the new spectrum is characterized by a new q_d given by translation law:

$$q_d = q + d. \qquad (2.4\text{-}10)$$

An optical system would usually comprise lenses and/or mirrors spaced apart from each other. While Gaussian beam propagation in between lenses and mirrors can be tracked using the translation law above, we need to develop the law of q -transformation by a lens. Note that the transparency function for a lens is of the form $\exp[jk_0(x^2+y^2)/2f]$. The optical field immediately behind the lens is therefore the product of the optical field immediately in front of the lens [see Eq. (2.4-6a)] and the transparency function, and can be expressed as

$$e^{-jk_0z}\,j\frac{k_0w_0^2}{2q}e^{-jk_0(x^2+y^2)/2q}e^{jk_0(x^2+y^2)/2f}$$

$$=e^{-jk_0z}\,j\frac{k_0w_0^2}{2q}e^{-jk_0(x^2+y^2)/2q_L}\,,$$

where q_L , the transformed q , is given by

$$\frac{1}{q_L}=\frac{1}{q}-\frac{1}{f}. \tag{2.4-11}$$

The laws of q-transformation [see Eqs. (2.4-10) and (2.4-11)] due to translation and lensing can be incorporated into a single relation using the *ABCD* parameters introduced in Chapter 1. The q-parameter transforms in general according to the *bilinear transformation*

$$q_2=\frac{Aq_1+B}{Cq_1+D}, \tag{2.4-12}$$

where the *ABCD* are the elements of the ray matrix transfer matrix which relates q_2 at output plane 2 to q_1 at input plane 1. For instance, the *ABCD* matrix for translation is $\begin{pmatrix}1 & d\\0 & 1\end{pmatrix}$, while that for a lens is $\begin{pmatrix}1 & 0\\-1/f & 1\end{pmatrix}$. Substitution of the values for A,B,C,D for each of the cases of translation and lensing gives the relations derived in Eqs. (2.4-10) and (2.4-11) above. In general, if we have two transformations

$$q_j = \frac{A_i q_i + B_i}{C_i q_i + D_i}, \qquad (2.4\text{-}13)$$

and

$$q_k = \frac{A_j q_j + B_j}{C_j q_j + D_j}, \qquad (2.4.14)$$

we can obtain

$$q_k = \frac{A_k q_k + B_k}{C_k q_k + D_k}, \qquad (2.4\text{-}15)$$

where

$$\begin{pmatrix} A_k & B_k \\ C_k & D_k \end{pmatrix} = \begin{pmatrix} A_j & B_j \\ C_j & D_j \end{pmatrix} \begin{pmatrix} A_i & B_i \\ C_i & D_i \end{pmatrix}.$$

We can verify Eq.(2.4-15) by substitution of Eq.(2.4-13) into Eq.(2.4-14). These q-transformations of Gaussian beams are Kogelnik's $ABCD$ law [Kogelnik (1965)].

Example: Focusing of a Gaussian beam

We analyze the focusing of a Gaussian beam by a positive lens. Assume that a Gaussian beam of initial waist w_0 [see Eq. (2.4-8), $R(0) = \infty$, and $w(0) = w_0$] , and correspondingly initial $q_0 = j z_R = j k_0 w_0^2 / 2$, is incident on a lens with a focal length f. Notice that the initial q is purely imaginary corresponding to a Gaussian beamwidth of initial planar wavefronts. After propagation through a distance z behind the lens, the ABCD matrix is

$$\begin{pmatrix} A & B \\ C & D \end{pmatrix} = \begin{pmatrix} 1 & z \\ 0 & 1 \end{pmatrix} \begin{pmatrix} 1 & 0 \\ -1/f & 1 \end{pmatrix} = \begin{pmatrix} 1 - z/f & z \\ -1/f & 1 \end{pmatrix},$$

and hence $q(z)$ of the beam is transformed to, according to Eq.(2.4-15),

$$q(z) = \frac{(1 - z/f)q_0 + z}{(-1/f)q_0 + 1} = \frac{f q_0}{f - q_0} + z. \qquad (2.4\text{-}16)$$

The Gaussian beam is said to be focused at the point $z = z_f$ where $q(z_f)$ becomes purely imaginary again or the Gaussian beam has a planar wavefront. Thus, setting $q(z_f) = jk_0 w_f^2 / 2$ in Eq.(2.4-16), we obtain

$$\left(z_f - \frac{f z_R^2}{f^2 + z_R^2} \right) + j \frac{f z_R}{f^2 + z_R^2} = jk_0 w_f^2 / 2,$$

where w_f is the waist at $z = z_f$. Equating the imaginary and real parts and simplifying, we have

$$z_f = \frac{f z_R^2}{f^2 + z_R^2} \tag{2.4-17}$$

and

$$w_f^2 = \frac{f^2 w_0^2}{f^2 + z_R^2}. \tag{2.4-18}$$

The Gaussian beam does not exactly focus at the geometrical back focus of the lens [Franco, Cywiak, Cywiak, and Mourad (2016)]. Instead, the focus is shifted closer to the lens, and tends to the geometrical focus at f as $w_0 \to \infty$, which is the case for a plane wave incidence. This effect is known as the *focal shift*, which has been conveniently analyzed based on the Fourier-optics approach [Poon (1987)]. For large w_0, we can show that

$$w_f \approx w_0 f / z_R = \lambda_0 f / \pi w_0, \tag{2.4-19}$$

where λ_0 is the optical wavelength. As an example, for $w_0 = 3$mm, $\lambda_0 = 0.633$ μm, and $f = 10$ cm, the focal spot size $w_f \approx 20$ μm.

2.4.2 *MATLAB example: Propagation of a Gaussian beam*

The following example shows the propagation of a Gaussian beam with initially planar phase-fronts in paraxial region. The MATLAB code that calculates the propagation of the Gaussian beam by the distance z is given below in Table 2.5. We have implemented the convolution process by Fourier transformation. Basically, the Fresnel diffraction formula is implemented by Eq. (2.3-14). Figures 2.14(a) and (b) show the initial and the diffracted beam profile, respectively. For numerical calculations that give Fig. 2.14, we input 0.6328 μm for the wavelength of HeNe laser. We in turn then input the size of initial waist $w_0 = 1\ mm$ and the program outputs the value of the Rayleigh range $z_R = 4964.590161\ mm$. The program also gives the peak amplitudes of the initial Gaussian beam and the diffracted Gaussian beam. When we input the value of the Rayleigh range, we have the peak amplitude of the diffracted beam drops to 0.692970, which is about $1/\sqrt{2}$ times its initial value of unity as expected. To increase the accuracy of the result, one can increase the sampling number N in the m-file to say, N = 300. We have used N = 100 in the calculations.

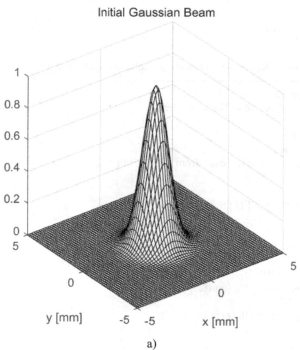

a)

Fig. 2.14 (a) Initial Gaussian and (b) after travel by a distance equal to the Rayleigh range of the initial Gaussian beam.

Propagated Gaussian Beam

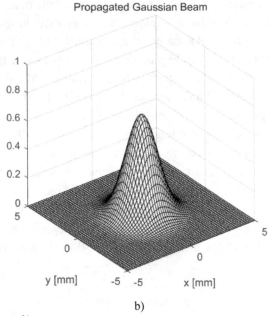

b)

Fig. 2.14 (*Continued*)

Table 2.5 Gaussian_propagation.m (m-file for calculating diffracted Gaussian beam).

```
%Gaussian_propagation.m
%Simulation of diffraction of Gaussian Beam
clear
%Gaussian Beam
%N : sampling number
N=input('Number of samples (enter from 100 to 500) = ');

L=10*10^-3;
Ld=input('wavelength of light in [micrometers] = ');
Ld=Ld*10^-6;
ko=(2*pi)/Ld;
wo=input('Waist of Gaussian Beam in [mm] = ');
wo=wo*10^-3;
z_ray=(ko*wo^2)/2*10^3;
sprintf('Rayleigh range is %f [mm]', z_ray)
z_ray=z_ray*10^-3;
z=input('Propagation length (z) in [mm] = ');
z=z*10^-3;

% dx : step size
dx=L/N;
```

```
for n=1:N+1
    for m=1:N+1

        %Space axis
        x(m)=(m-1)*dx-L/2;
                    y(n)=(n-1)*dx-L/2;

        % Gaussian Beam in space domain
                    Gau(n,m)=exp(-(x(m)^2+y(n)^2)/(wo^2));

        %Frequency axis
        Kx(m)=(2*pi*(m-1))/(N*dx)-((2*pi*(N))/(N*dx))/2;
        Ky(n)=(2*pi*(n-1))/(N*dx)-((2*pi*(N))/(N*dx))/2;

        %Free space transfer function
        H(n,m)=exp(j/(2*ko)*z*(Kx(m)^2+Ky(n)^2));
            end
end

%Gaussian Beam in Frequency domain
FGau=fft2(Gau);
FGau=fftshift(FGau);

%Propagated Gaussian beam in Frequency domain
FGau_pro=FGau.*H;

%Peak amplitude of the initial Gaussian beam
Peak_ini=max(max(abs(Gau)));
sprintf('Initial peak amplitude is %f [mm]', Peak_ini)
%Propagated Gaussian beam in space domain
Gau_pro=ifft2(FGau_pro);
Gau_pro=Gau_pro;

%Peak amplitude of the propagated Gaussian beam
Peak_pro=max(max(abs(Gau_pro)));
sprintf('Propagated peak amplitude is %f [mm]', Peak_pro)

%Calculated Beam Width
[N M]=min(abs(x));
Gau_pro1=Gau_pro(:,M);
[N1 M1]= min(abs(abs(Gau_pro1)- abs( exp(-1)*Peak_pro)));
Bw=dx*abs(M1-M)*10^3;
sprintf('Beam width (numerical)is %f [mm]', Bw)
%Theoretical Beam Width
W=(2*z_ray)/ko*(1+(z/z_ray)^2);
W=(W^0.5)*10^3;
sprintf('Beam width (theoretical)is %f [mm]', W)
```

```
%axis in mm scale
x=x*10^3;
y=y*10^3;

figure(1);
mesh(x,y,abs(Gau))
title('Initial Gaussian Beam')
xlabel('x [mm]')
ylabel('y [mm]')
axis([min(x) max(x) min(y) max(y) 0 1])
axis square

figure(2);
mesh(x,y,abs(Gau_pro))
title('Propagated Gaussian Beam')
xlabel('x [mm]')
ylabel('y [mm]')
axis([min(x) max(x) min(y) max(y) 0 1])
axis square
```

Problems:

2.1   Show that the wave equation for $\mathcal{H}$ in a linear, homogeneous, and isotropic medium characterized by μ and ε is given by

$$\nabla^2 \mathcal{H} - \mu\varepsilon \frac{\partial^2 \mathcal{H}}{\partial t^2} = -\nabla \times \boldsymbol{J}.$$

2.2   Determine which of the following functions describe traveling waves (a and b are some real constant).

a) $\psi(z,t) = e^{-(a^2 z^2 + b^2 t^2 + 2abzt)}$

b) $\psi(z,t) = \cos[(az - bt)(az + bt)]$

c) $\psi(z,t) = \sin^3[(\frac{z}{a} + \frac{t}{b})^2]$

d) $\psi(z,t) = \mathrm{sech}(at - bz)$

2.3 Show that $\psi(r,t) = J_0(k_0 r)e^{j\omega_0 t}$ is an exact solution to the 3-D scalar wave equation [see Eq. (2.2-18)] under the assumption of cylindrical symmetry, where $u = \omega_0 / k_0$ and $J_0(.)$ is Bessel's function of zero order. When $r \gg 1$, verify that

$$\psi(r,t) \to \frac{1}{\sqrt{r}} e^{j(\omega_0 t - k_0 r)}.$$

2.4 Show that $\psi(z,t) = c_1 f(\omega_0 t - k_0 z) + c_2 g(\omega_0 t + k_0 z)$, where $u = \omega_0 / k_0$, f and g are arbitrary functions, is a general solution to the 1-D scalar wave equation given by Eq. (2.2-18).

2.5 Derive the amplitude reflection and transmission coefficients given by Eq. (2.2-47) for perpendicular polarization.

2.6 Verify Eqs.(2.2-47a) and (2.2-47b), and plot the incident angle as a function of the phase α of the reflected wave in the case of perpendicular polarization for glass to air index of refraction ratio of 1.5, which is shown in Fig. 2.5(b). Your plot should confirm the equation for α given by Eq. (2.2-58).

2.7 Verify the Fourier transform pairs 6 and 10 in Table 2.2.

2.8 Show that the operation of convolution is commutative, i.e.,

$$g_1(x,y) * g_2(x,y) = g_2(x,y) * g_1(x,y).$$

2.9 Show that

$$\mathcal{F}_{xy}\{g_1(x,y) * g_1(x,y)\} = G_1(k_x,k_y) G_2(k_x,k_y)$$

where $G_1(k_x,k_y)$ and $G_2(k_x,k_y)$ are the Fourier transforms of $g_1(x,y)$ and $g_2(x,y)$, respectively.

2.10   Find a paraxial approximation to a wavefront, in the plane $z = 0$
, that converges to the point P as shown in Fig. P2.10. A
transparency $t(x, y)$ is now placed at $z = 0$ and illuminated by the
converging wavefront. Assuming Fresnel diffraction from $z = 0$
to the plane $z = z_0$, find the complex field on the observation
plane. Comment on the usefulness of illuminating an
transparency using a spherical wavefront instead of a plane
wavefront. [Adopted from Banerjee and Poon, Principles of
Applied Optics, Irwin, 1991].

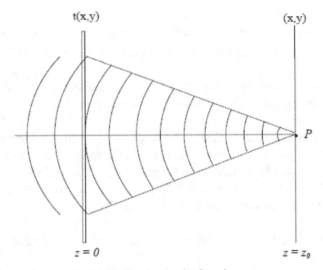

Fig. P2.10 Illumination by focusing wave.

2.11   A *sinusoidal amplitude grating* given by

$$t(x, y) = \left[\frac{1}{2} + \frac{m}{2} \cos(ax) \right] rect\left(\frac{x}{l}, \frac{y}{l} \right), m < 1$$

is illuminated by a plane wave of unit amplitude. Determine its
Fraunhofer diffraction. Plot the intensity distribution along the
x-axis and label all essential points along the axis.

2.12   A double-slit described by the transparency function

$$t(x,y) = rect\left[\frac{x - X/2}{x_0}\right] + rect\left[\frac{x + X/2}{x_0}\right], X >> x_0$$

is illuminated by a plane wave of unit amplitude. Determine its Fraunhofer diffraction. Plot the intensity distribution along the x-axis and label all essential points along the axis.

2.13   A plane wave of unit amplitude propagating in the $+z$ direction is incident normally on an infinite series of slits, spaced S apart and a wide at $z = 0$, as shown in Fig. P2.13. Such a grating is called a *Ronchi grating*. Under Fraunhofer diffraction of the grating, find an expression for the complex amplitude. Sketch the intensity distribution on the observation plane, labeling any coordinates of important points [Adopted from Banerjee and Poon, Principles of Applied Optics, Irwin, 1991].

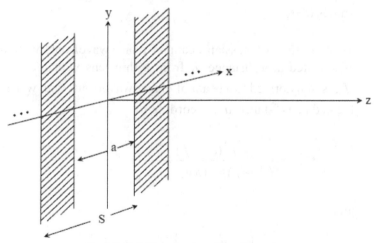

Fig. P2.13 Diffraction by a Ronchi grating.

2.14   Find the Fraunhofer diffraction pattern of a *sinusoidal phase grating* described by $\exp[j(m/2)\sin(\alpha x)]rect(x/l)rect(y/l)$, where $l >> 2\pi/\alpha$. Describe qualitatively what you would expect if the phase grating is moving along x with a velocity

$V = \Omega / a$, a situation encountered in acousto-optics to be discussed in Chapter 4 [Adopted from Banerjee and Poon, Principles of Applied Optics, Irwin, 1991].

2.15   For a given pupil function $p(x, y)$ in the 4-f optical system, its coherent function is given by

$$h_c(x, y) = \mathcal{F}_{x,y}\{p(x, y)\}\Big|_{kx=k_0x/\,f, ky=k_0y/\,f} = P\left(\frac{k_0 x}{f}, \frac{k_0 y}{f}\right).$$

Show that the coherent transfer function is

$$H_c\left(k_x, k_y\right) = \mathcal{F}_{x,y}\{h_c(x, y)\} = p\left(\frac{-fk_x}{k_0}, \frac{-fk_y}{k_0}\right).$$

2.16   For a given pupil function $p(x, y) = 1 - circ(r / r_0)$ in the 4-f optical system, write a MATLAB program to compute and display the filtered intensity image on the image plane for various values of r_0.

2.17   a) Show that a Gaussian beam of plane wavefronts with waist w_1, located at a distance d_1 from a thin lens with focal length f, is transformed to a beam of planar wavefronts with waist w_2, located at a distance d_2, according to

$$d_2 = \frac{f^2(d_1 - f)}{(d_1 - f)^2 + (\pi w_1 / \lambda_0)^2} + f$$

and

$$w_2 = \left[\frac{1}{w_1^2}\left(1 - \frac{d_1}{f}\right)^2 + \frac{1}{f^2}\left(\pi w_1 / \lambda_0\right)^2\right]^{-1/2}.$$

The situation is shown in Fig. P2.17.

b) When $w_1 = 0$, which celebrated formula have you recovered?

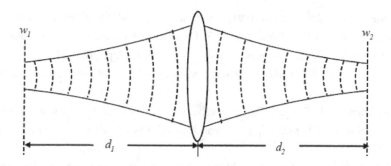

Fig. P2.17 Imaging of a Gaussian beam.

2.18 Propagation of a Gaussian beam through the square-law medium $n^2(x,y) = n_0^2 - n_2(x^2 + y^2)$ with $\beta = \sqrt{n_2} / n_0$:

a) Assume that the Gaussian beam of wavelength λ_0 at $z = 0$ in air has waist w_0, show that the beam width at z, i.e., $w(z)$, is given by

$$w(z) = w_0 \left[1 + \frac{\lambda_0^2 - \pi^2 \beta^2 n_0^2 w_0^4}{\pi^2 \beta^2 n_0^2 w_0^4} \sin^2 \beta z \right]^{\frac{1}{2}}$$

using the q-transformation.

b) Show that, unless $\lambda_0 = \pi \beta n_0 w_0^2$ or the incident waist satisfies $w_0^2 = \lambda_0 / \pi \sqrt{n_2}$, the width $w(z)$ fluctuates between extreme values. Find these extreme values.

c) Plot $w(z)$ vs. z for the following cases:

1) $\pi \beta n_0 w_0^2 > \lambda_0$; 2) $\pi \beta n_0 w_0^2 < \lambda_0$,

and explain what is the difference between the two cases.

Bibliography

Banerjee, P. P. and T.-C. Poon (1991). *Principles of Applied Optics*. Irwin, Illinois.

Banerjee, P. P. and T.-C. Poon (1995). "Self-Fourier Objects and Other Self-Transform Objects: Additional Comments," *J. Opt. Soc. Am. A*, 12, pp. 425-426.

Cheng, D. K. (1983). *Field and Wave Electromagnetics*. Addison-Wesley, Reading, Massachusetts.

J. M. Franco, M. Cywiak, D. Cywiak, and I. Mourad (2016). "Optical Focusing Conditions of Lenses using Gaussian Beams," *Optics Communications*, 371, pp. 226–230.

Ghatak, A. K. and K. Thyagarajan (1989). *Optical Electronics*, Cambridge University Press, Cambridge.

Goodman, J. W. (1996). *Introduction to Fourier Optics*. McGraw-Hill, New York.

Guenther, R. D. (1990). *Modem Optics*. John Wiley & Sons, New York.

Hecht, E. and A. Zajac (1975). *Optics*. Addison-Wesley, Reading, Massachusetts.

Kogelnik, H. (1965). "Imaging of Optical Mode Resonators with Internal Lenses," *Bell Syst. Tech. J.*, 44, 455-494.

Marcuse, D. (1982). *Light Transmission Optics*. Van Nostrand Reinhold Company, New York.

Poon, T.-C and M. Motamedi (1987). "Optical/Digital Incoherent Image Processing for Extended Depth of Field," *Appl. Opt.*, 26, pp. 4612-4615.

Poon, T.-C. (1988). "Focal Shift in Focused Annular Beams," *Optics Communications*, 6, pp. 401-406.

Poon T.-C. and P. P. Banerjee (2001). *Contemporary Optical Image Processing with MATLAB®*. Elsevier, Oxford, UK.

Poon T.-C. (2007). *Optical Scanning Holography with MATLAB®*. Springer, New York.

Poon, T.-C. and J.-P. Liu (2014). *Introduction to Modern Digital Holography with MATLAB*, Cambridge University Press, Cambridge, UK.

Slepian, D. and H.O. Pollak (1961). "Prolate Spheroidal Wave Functions, Fourier Analysis and Uncertainty - I," *Bell Syst. Tech. J.,* 40, pp. 43-63.

Stark, H., ed. (1982). *Applications of Optical Transforms*. Academic Press, Florida.
Ulaby, F.T. (2002). *Fundamentals of Applied Electromagnetics*, Prentice-Hall.

Chapter 3

Beam Propagation in Inhomogeneous Media and in Kerr Media

3.1 Wave Propagation in a Linear Inhomogeneous Medium

Thus far, as in Chapter 2, we have only considered wave propagation in a homogeneous medium, characterized by a constant permittivity ε. In inhomogeneous materials, the permittivity can be a function of the spatial coordinates x, y and z, i.e., $\varepsilon(x,y,z)$. To study wave propagation in inhomogeneous materials, we return to Maxwell's equations (2.1-1)-(2.1-4) and re-derive the wave equation. Our starting point is Eq. (2.2-4) which we rewrite here for $\boldsymbol{J} = 0$:

$$\nabla^2 \boldsymbol{E} - \mu\varepsilon\frac{\partial^2 \boldsymbol{E}}{\partial t^2} = \nabla\left(\nabla\cdot\boldsymbol{E}\right). \tag{3.1-1}$$

Now, from Eq. (2.1-1) with $\rho_v = 0$, and Eq. (2.1-12a), we have

$$\nabla\cdot(\varepsilon\boldsymbol{E}) = \varepsilon\nabla\cdot\boldsymbol{E} + \boldsymbol{E}\cdot\nabla\varepsilon = 0. \tag{3.1-2}$$

Substituting Eq. (3.1-2) into Eq. (3.1-1) yields

$$\nabla^2 \boldsymbol{E} - \mu\varepsilon\frac{\partial^2 \boldsymbol{E}}{\partial t^2} = -\nabla\left(\boldsymbol{E}\cdot\frac{\nabla\varepsilon}{\varepsilon}\right). \tag{3.1-3}$$

The right-hand side of the above equation is in general non-zero when there is a gradient in the permittivity of the medium, such as the case in guided-wave optics. However, if the spatial variation of the refractive index is small over the distance of one optical wavelength, the term $\nabla \varepsilon / \varepsilon \approx 0$ [Marcuse, 1982]. This approximation is of a similar nature as the paraxial approximation. If we were content with this approximation, we can study the propagation of light in inhomogeneous media by neglecting the right-hand side of Eq. (3.1-3), giving us the following wave equation to solve:

$$\nabla^2 \boldsymbol{E} - \mu\varepsilon \frac{\partial^2 \boldsymbol{E}}{\partial t^2} = 0, \tag{3.1-4}$$

where $\varepsilon = \varepsilon(x, y, z)$. Note that Eq. (3.1-4) is similar to the homogeneous wave equation for the electric field Eq. (2.2-10) derived earlier. For notational convenience, we return to our generic dependent variable $\psi(x, y, z, t)$ and adopt

$$\nabla^2 \psi - \mu\varepsilon \frac{\partial^2 \psi}{\partial t^2} = 0, \quad \varepsilon = \varepsilon(x, y, z) \tag{3.1-5}$$

as our model equation, where we have assumed $\mu = \mu_0$ for simplicity. Again, ψ may represent a component of the electric field, $\boldsymbol{E}$.

3.2 Optical Propagation in Square-Law Media

A square-law medium with an index of refraction of the form $n^2(x, y) = n_0^2 - n_2(x^2 + y^2)$ has been studied in Chapter 1 using ray optics. Equivalently, we can incorporate the inhomogeniety through a square-law permittivity profile of the form [Haus (1984)]

$$\varepsilon(x, y, z) = \varepsilon(x, y) = \varepsilon_0 \varepsilon_r(x, y) = \varepsilon(0)\left(1 - \frac{x^2 + y^2}{h^2}\right). \tag{3.2-1}$$

We wish to study the propagation of arbitrary beam profiles through the inhomogeneous medium modeled by Eq. (3.2-1). However, analytical solutions of Eq. (3.1-5) with Eq. (3.2-1) is difficult for some arbitrary initial conditions. We will, therefore, first look for a propagating plane wave solution which can have an arbitrary cross-sectional amplitude and/or phase profile. Thus, we let

$$\psi(x,y,z,t) = \psi_p(x,y,z)\exp\left[j(\omega_0 t)\right] = \psi_e(x,y)\exp\left[j(\omega_0 t - \beta z)\right],$$
(3.2-2)

where $\psi_e(x,y)$ is called the *complex envelope*, and β is the *propagation constant* as yet to be determined. We have assumed that $\psi_e(x,y)$ is not a function of z for the reason that $\varepsilon(x,y,z) = \varepsilon(x,y)$, which is not a function of z. By substituting Eq. (3.2-2) into the wave equation, Eq. (3.1-5), we have

$$\nabla_t^2 \psi_e + \left[\omega_0^2 \mu_0 \varepsilon(x,y) - \beta^2\right]\psi_e = 0,$$
(3.2-3)

where ∇_t^2 denotes the transverse Laplacian $\frac{\partial^2}{\partial x^2} + \frac{\partial^2}{\partial y^2}$. We denote by k_0 the propagation constant of an infinite plane wave propagating in a medium of uniform dielectric constant $\varepsilon(0)$, i.e.,

$$k_0 = \omega_0 \left[\mu_0 \varepsilon(0)\right]^{1/2}.$$
(3.2-4)

When we introduce Eq. (3.2-1) into Eq. (3.2-3) with Eq. (3.2-4) and use the normalized variables

$$\xi = \gamma x \text{ and } \eta = \gamma y ,$$
(3.2-5)

where

$$\gamma = \left(\frac{k_0}{h}\right)^{1/2},$$

we get

$$\frac{\partial^2 \hat{\psi}_e(\xi,\eta)}{\partial \xi^2} + \frac{\partial^2 \hat{\psi}_e(\xi,\eta)}{\partial \eta^2} + \left[\lambda - (\xi^2 + \eta^2)\right]\hat{\psi}_e(\xi,\eta) = 0,$$
(3.2-6a)

where

$$\lambda = \frac{\left(k_0^2 - \beta^2\right)h}{k_0}. \tag{3.2-6b}$$

We solve Eq. (3.2-6) using the commonly used *separation of variables* technique. We assume $\hat{\psi}_e(\xi,\eta) = X(\xi)Y(\eta)$, substitute it in Eq. (3.2-6) and derive two decoupled ordinary differential equations (ODEs) for X and Y. These are

$$\frac{d^2 X}{d\xi^2} + \left(\lambda_x - \xi^2\right)X = 0 \tag{3.2-7a}$$

and

$$\frac{d^2 Y}{d\eta^2} + \left(\lambda_y - \eta^2\right)Y = 0 \tag{3.2-7b}$$

with $\lambda_x + \lambda_y = \lambda$. Each of Eqs. (3.2-7a) and (3.2-7b) is of the same form as that arising in the analysis of the harmonic oscillator problem in quantum mechanics [Schiff (1968)]. The solution to Eq. (3.2-7a) is

$$X_m(\xi) = H_m(\xi)\exp\left[-\xi^2/2\right], \lambda_x = 2m+1; \ m = 0,1,2,\ldots,$$
$$\tag{3.2-8}$$

where the H_ms are the *Hermite polynomials* defined by Eq. (2.3-35a). Again, the first three Hermite polynomials are

$$H_0(\xi) = 1, \ H_1(\xi) = 2\xi; \ H_2(\xi) = 4\xi^2 - 2. \tag{3.2-9}$$

The solutions to $X(\xi)$ are *Hermite-Gaussians*. The first few have been plotted in Figure 2.11. Similar solutions hold for $Y(\eta)$. Equation (3.2-6) thus has the general solution

$$\hat{\psi}_e(\xi,\eta) = \hat{\psi}_{emn}(\xi,\eta) = H_m(\xi)H_n(\eta)\exp\left[-\left(\xi^2 + \eta^2\right)/2\right]$$
$$\tag{3.2-10a}$$

with

$$\lambda = \lambda_{mn} = 2(m+n+1); \ m,n = 0,1,2,\ldots \tag{3.2-10b}$$

$\hat{\psi}_{emn}$ is called the *mode pattern* or *mode profile* of the *mn*-th mode. The above analysis is indicative of the mode patterns that are characteristic of a multimode graded-index optical fiber. An arbitrary excitation at the input of the fiber can be tracked by decomposing it into the characteristic modes discussed above. The Hermite-Gaussian functions form an orthogonal basis, enabling such a decomposition to be made easily. To formulate the idea just mentioned, we write the normalized *mn*-th mode patterns as

$$\hat{\psi}_{emn}(x,y) = u_m(x)u_n(y), \qquad (3.2\text{-}11\text{a})$$

where

$$u_m(x) = N_m H_m(\xi)\exp\left[-\xi^2/2\right] \qquad (3.2\text{-}11\text{b})$$

and

$$u_n(y) = N_n H_n(\eta)\exp\left[-\eta^2/2\right] \qquad (3.2\text{-}11\text{c})$$

are the normalized functions and

$$N_p = \left(\frac{\gamma}{\sqrt{\pi 2^p \, p!}}\right)^{\frac{1}{2}}, \, p = m \text{ or } n$$

is the normalization constant such that

$$\int_{-\infty}^{\infty}\int_{-\infty}^{\infty}\psi_{emn}(x,y)\,\psi_{emn}^{*}(x,y)\,dxdy = \delta_{mn} = \begin{cases} 1 & m = n \\ 0 & m \neq n \end{cases} \qquad (3.2\text{-}12)$$

and the corresponding propagation constants $\beta = \beta_{mn}$, from Eq. (3.2-6b) with Eq. (3.2-10b), are

$$\beta_{mn}^2 = k_0^2\left(1 - \frac{2(m+n+1)}{k_0 h}\right). \qquad (3.2\text{-}13)$$

As an example, the *fundamental mode*, with $m = n = 0$ $(\lambda_{00} = 2)$ is given by

$$\psi_{e00}(x,y) = \frac{(2/\pi)^{1/2}}{w_{00}} \exp\left[-\left(x^2 + y^2\right)/w_{00}^2\right], \quad (3.2\text{-}14a)$$

where

$$w_{00} = \sqrt{2}/\gamma. \qquad (3.2\text{-}14b)$$

The function $\psi_{e00}(x,y)$ forms a complete set of orthonormal functions and the propagation of a beam in a square-law medium can then be written in the form of, according to Eq. (3.2-2),

$$\psi_p(x,y,z) = \sum_m \sum_n c_{mn} u_m(x) u_n(y) \exp\left[-j\beta_{mn}z\right], \quad (3.2\text{-}15)$$

where c_{mn} is a constant that is determined from the knowledge of $\psi_p(x,y,0)$ and the use of Eq. (3.2-12):

$$c_{mn} = \int_{-\infty}^{\infty}\int_{-\infty}^{\infty} \psi_p(x',y',0)u_m^*(x')u_n^*(y')dx'dy'. \quad (3.2\text{-}16)$$

Hence for a given incident field $\psi_p(x,y,0)$ to the square-law medium, the field distribution at an arbitrary distance z is given by substituting Eq. (3.2-16) into Eq. (3.2-15):

$$\psi_p(x,y,z) = \int_{-\infty}^{\infty}\int_{-\infty}^{\infty} K(x,y,x',y')\psi_p(x',y',0)dx'dy', \quad (3.2\text{-}17)$$

where

$$K(x,y,x',y') = \sum_m \sum_n u_m^*(x)u_m(x')u_n^*(y)u_n(y')\exp\left[-j\beta_{mn}z\right].$$

The above equation can be summed to give an analytical expression if we make an approximation on β_{mn} such that k_0h is large (which is the case for practical graded-index fibers). Eq. (3.2-13) becomes

$$\beta_{mn} \approx k_0 - \frac{(m+n+1)}{h}. \qquad (3.2\text{-}18)$$

With the above approximation and after the summations, $K(x, y, z', y')$ in Eq. (3.2-17) becomes [Sodha and Ghatak, 1977]

$$K(x, y, x', y') = \frac{j\gamma^2}{2\pi \sin(z/h)} \exp[-jk_o z]$$

$$\times \exp\left[\frac{j\gamma^2}{\sin(z/h)}(xx' + yy') - j\frac{\gamma^2}{2}\left(x^2 + x'^2 + y^2 + y'^2\right)\cot(z/h)\right].$$

$$(3.2\text{-}19)$$

Now, let us return to the discussion on β_{mn}. Using Eq. (3.2-4) and from Eq. (3.2-18), we have

$$\beta_{mn} \approx \omega_0 [\mu_0 \varepsilon(0)]^{1/2} - \frac{(m+n+1)}{h}.$$

If we define $[\mu_0 \varepsilon(0)]^{1/2} = \sqrt{\varepsilon_r} / c$ and neglect *material dispersion*, i.e., the frequency dependence of ε_r, we then have

$$\frac{d\beta_{mn}}{d\omega_0} = \frac{\sqrt{\varepsilon_r}}{c}.$$

Therefore, the *group velocity*, u_g, of the *mn*-th mode is

$$u_g = \frac{d\omega_0}{d\beta_{mn}} = \frac{c}{\sqrt{\varepsilon_r}}, \qquad (3.2\text{-}20)$$

which is independent of the mode numbers *m* and *n*. In other words, different modes in a square-law medium travel with the same group velocity. This is a remarkable result. Unlike step-index fibers shown in Fig. 3.1, where a uniform refractive index n_1 is surrounded by a cladding of another material of uniform but slightly lower refractive index n_2. Step-index fibers have the so-called *intermodal dispersion*. From the definition of the critical angle given by Eq. (1.2-6) and with reference to Fig. 3.1, we see that for a guided ray, we must satisfy

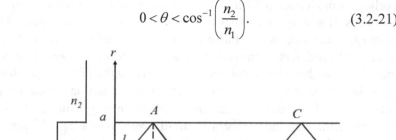

$$0 < \theta < \cos^{-1}\left(\frac{n_2}{n_1}\right). \qquad (3.2\text{-}21)$$

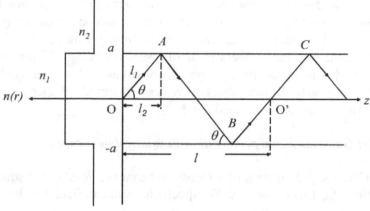

Fig. 3.1 Ray diagram of a step-index fiber.

Since the group velocities of all rays here are the same, we see that the ray traveling along the axis of the fiber, i.e., $\theta = 0$, will travel shorter distance than the ones traveling along the zigzagged paths. Indeed, the ray traveling along the zigzagged path that corresponds to $\theta = \theta_c$ (the critical angle) will travel the longest distance, and hence take the longest time among the many modes to complete one cycle for a distance from O to O'. The time for the ray to travel along the axis is $t_a = l/(c/n_1)$, and the time for the critical-angle ray is $t_c = 4l_1/(c/n_1)$. Now from geometry, $l_1 = l_2/\cos\theta_c$, and $\cos\theta_c = n_2/n_1$, we then have $t_c = l\,n_1^2/(cn_2)$ using $4l_2 = l$. Hence if all the input rays were excited at the same time, the rays will span a time interval at the output end with duration

$$\Delta t = (t_c - t_a) = \frac{n_1(n_1 - n_2)l}{cn_2}. \qquad (3.2\text{-}22)$$

For a typical fiber, if we take $n_1 = 1.46$, $(n_1 - n_2)/n_2 = 0.01$, and $l = 1$ *km*, we have $\Delta t \approx 50ns/km$. In other words, an impulse after traveling through such a fiber of 1 km long would broaden to a pulse of about 50ns. Hence in a 1 Gbit/s fiber optics communication system, where the transmission rate is one pulse every $10^{-9}s$, a dispersion of 50ns/km would result in detector errors for distances exceeding about 20m if we want to resolve these individual pulses. To reduce the pulse dispersion for high information-carrying system, we could use square-law medium-type fibers or employ the use of single mode fibers. In single mode fibers, the core size of the step index fibers is small (usually less than 10 μm) and only one mode (the mode that propagates straight along the fiber axis) is allowed to propagate. In such fibers, there is no intermodal dispersion but material dispersion has to be considered.

Example: Gaussian beam propagation in square-law medium

We consider the propagation of a Gaussian beam incident on the square law medium, Eq. (3.2-1), at $z = 0$. We specify the incident field distribution as

$$\psi_p(x,y;z=0) = \psi_0 \exp\left[-\frac{x^2+y^2}{w_0^2}\right]. \qquad (3.2\text{-}23)$$

Substituting Eqs. (3.2-23) and (3.2-19) into Eq. (3-2-17), we have the field distribution in any transverse plane as [Ghatak and Thyagarajan (1978)]

$$\psi_p(x,y,z) = \frac{w_0\psi_0 e^{-j\varphi(z)}}{w(z)} \exp\left[-\frac{x^2+y^2}{w^2(z)}\right], \qquad (3.2\text{-}24)$$

where

$$w^2(z) = \frac{4}{w_0^2\gamma^4}\left[\sin^2\left(\frac{z}{h}\right) + \frac{w_0^4\gamma^4}{4}\cos^2\left(\frac{z}{h}\right)\right],$$

and

$$\varphi(z) = k_0 z - \tan^{-1}\left(\frac{\tan(z/h)}{\tau^2}\right) - \frac{(\tau^4-1)\sin(2z/h)}{2\tau^2 w^2(z)}(x^2+y^2)$$

with $\tau = w_0\gamma/\sqrt{2}$ .Eq. (3.2-24) indicates that as the Gaussian beam propagates in the square-law medium, it always remains Gaussian with its waist modulating with a period $z_m = \pi h$. In Section 3.5, we will demonstrate such modulation effect of the beam waist using numerical methods.

3.3 The Paraxial Wave Equation

Let us return to the model equation given by Eq. (3.1-5) for inhomogeneous media:

$$\nabla^2 \psi - \mu\varepsilon \frac{\partial^2 \psi}{\partial t^2} = 0, \quad \varepsilon = \varepsilon(x,y,z). \tag{3.3-1}$$

If we now write

$$\psi(x,y,z,t) = \psi_p(x,y,z)\exp[j(\omega_0 t)]$$
$$= \psi_e(x,y,z)\exp[j(\omega_0 t - k_0 z)], \tag{3.3-2}$$

we basically assume that ψ is predominantly propagating along the z-direction with complex envelope ψ_e. Note that we have assumed ψ_e is a function of z here, which is different from that assumed in Eq. (3.2-2). The reason is that we expect a change of the index refraction along z as indicated in general in Eq. (3.3-1). By substituting Eq. (3.3-2) into Eq. (3.3-1), we derive the following equation for the complex envelop:

$$\nabla_t^2 \psi_e + \frac{\partial^2 \psi_e}{\partial z^2} - 2jk_0 \frac{\partial \psi_e}{\partial z} - (k_0^2 - \mu_0\varepsilon\omega_0^2)\psi_e = 0. \tag{3.3-3}$$

We now assume that ψ_e is a slowly varying function of z in the sense that

$$\partial\psi_e/\partial z \ll k_0\psi_e. \tag{3.3-4}$$

This assumption physically means that within a wavelength of the propagation distance, i.e., $\Delta z \approx \lambda_0$, the change in ψ_e is much smaller than ψ_e itself, i.e., $\Delta\psi_e \ll \psi_e$. In differential form, $\Delta\psi_e \ll \psi_e$ becomes

$$d\psi_e = (\partial\psi_e/\partial z)\Delta z$$
$$= (\partial\psi_e/\partial z)\lambda_0 \ll \psi_e,$$

leading to the result in Eq. (3.3-4). Similarly, the derivative $\partial \psi_e / \partial z$ varies slowly within a distance of the wavelength so that $\partial^2 \psi_e / \partial z^2 << k_0 \partial \psi_e / \partial z$. Hence we can say that the contribution of the second term in Eq. (3.3-3) can be ignored when compared to the third term of the equation. Equation (3.3-3) then becomes the *paraxial Helmholtz equation* for ψ_e in inhomogeneous media, which reads

$$\nabla_t^2 \psi_e - 2jk_0 \frac{\partial \psi_e}{\partial z} - (k_0^2 - \mu_0 \varepsilon \omega_0^2) \psi_e = 0. \tag{3.3-5}$$

Note that for the special case of a homogenous medium, i.e., $\varepsilon(x,y) =$ constant $= \varepsilon(0)$, Eq. (3.3-5) is simplified to

$$\frac{\partial \psi}{\partial z} = \frac{1}{2jk_0} \nabla_t^2 \psi_e. \tag{3.3-6}$$

Fourier transforming Eq. (3.3-6) with respect to the variables x and y leads to the ordinary differential equation (ODE):

$$\frac{\partial \Psi_e}{\partial z} = -\frac{j(k_x^2 + k_y^2)}{2jk_0} \Psi_e, \tag{3.3-7}$$

where $\mathcal{F}_{xy}\{\psi_e(x,y,z)\} = \Psi_e(k_x,k_y;z)$. When Eq. (3.3-7) is solved for some initial field profile spectrum $\Psi_e(k_x,k_y;0)$, it yields a paraxial transfer function $H_e(k_x,k_y;z)$:

$$H_e(k_x,k_y;z) = \frac{\Psi_e(k_x,k_y;z)}{\Psi_e(k_x,k_y;0)} = \exp\left[\frac{j(k_x^2 + k_y^2)z}{2k_0}\right], \tag{3.3-8}$$

which is identical to the spatial frequency transfer function in Fourier optics given by Eq. (2.3-13) if we incorporate the term $\exp(-jk_0 z)$ from Eq. (3.3-2). Hence, the solution to Eq. (3.3-6) describes Fresnel diffraction for a given $\psi_e(x,y,0)$ in a homogeneous medium. Let us now return to Eq. (3.3-5) and inspect the position-dependent term $k_0^2 - \mu_0 \varepsilon \omega_0^2$. If we model

$$\mu_0 \varepsilon \omega_0^2 = \mu_0 \varepsilon_0 \varepsilon_r (1 + \Delta\varepsilon) \omega_0^2 = k_0^2 + k_0^2 \Delta\varepsilon,$$

where $\Delta\varepsilon = \Delta\varepsilon(x, y)$ and hence, the term $k_0^2 - \mu_0 \varepsilon \omega_0^2 = -k_0^2 \Delta\varepsilon$. Now we can write

$$n(x, y) = n_0(1 + \Delta n(x, y)) = \sqrt{\varepsilon_r (1 + \Delta\varepsilon)} \approx \sqrt{\varepsilon_r} + \frac{\sqrt{\varepsilon_r}}{2} \Delta\varepsilon.$$

We then see that $\Delta n(x, y) = \Delta\varepsilon/2$. So finally the term, $k_0^2 - \mu_0 \varepsilon \omega_0^2 = -k_0^2 \Delta\varepsilon = -k_0^2 2\Delta n$. Using this result, we can rewrite Eq. (3.3-5) to read

$$\frac{\partial \psi_e}{\partial z} = \frac{1}{2jk_0} \nabla_t^2 \psi_e - j\Delta n k_0 \psi_e. \tag{3.3-9}$$

The quantity Δn is the change in the refractive index over the ambient refractive index n_0. The above equation is called the *paraxial propagation equation*, which is a partial differential equation (PDE) that does not always lend itself to analytical solutions, except for some very special cases involving special spatial variations of Δn or when as in nonlinear optics, one looks for particular *soliton* solution of the resulting nonlinear partial differential equation using exact integration or inverse scattering methods. Numerical approaches are often sought for to analyze beam (and pulse) propagation in complex systems such as optical fibers, volume diffraction gratings, Kerr media, etc. A large number of numerical methods can be used for this purpose. The pseudospectral methods are often favored over *finite difference methods* due to their speed advantage. The *split-step beam propagation method* is an example of a pseudospectral method. We discuss this numerical technique in the next section.

3.4 The Split-Step Beam Propagation Method

To understand the philosophy behind the split-step beam propagation method, also called simply the *beam propagation method* (BPM), we re-write Eq. (3.3-9) in the operator-form [Agrawal, 1989]

$$\partial \psi_e / \partial z = (\hat{D} + \hat{S})\psi_e \, , \qquad (3.4\text{-}1)$$

where $\hat{D} = \frac{1}{2jk_0}\nabla_t^2$ is the linear differential operator that accounts for diffraction, also called the *diffraction operator*, and $\hat{S} = -j\Delta nk_0$ is the space-dependent or *inhomogeneous operator*. In general, the operators act together on ψ_e simultaneously and the operator-form solution of Eq. (3.4-1) is given by

$$\psi_e(x, y; z + \Delta z) = \exp[(\hat{D} + \hat{S})\Delta z]\psi_e(x, y; z) \qquad (3.4\text{-}2)$$

if $\hat{D}$ and $\hat{S}$ are z-independent. In general, for two commuting operators on $\psi_e(x, y; z)$, i.e., $[\hat{D}, \hat{S}]\psi_e = (\hat{D}\hat{S} - \hat{S}\hat{D})\psi_e = 0$, we have

$$\exp(\hat{D}\Delta z)\exp(\hat{S}\Delta z) = \exp[\hat{D}\Delta z + \hat{S}\Delta z].$$

However, the above equality does not necessarily hold for operators that do not commute. For two noncommuting operators $\hat{D}$ and $\hat{S}$, i.e., $[\hat{D}, \hat{S}]\psi_e = (\hat{D}\hat{S} - \hat{S}\hat{D})\psi_e \neq 0$, we have

$$\exp(\hat{D}\Delta z)\exp(\hat{S}\Delta z) = \exp[\hat{D}\Delta z + \hat{S}\Delta z + \frac{1}{2}[\hat{D}, \hat{S}](\Delta z)^2 + \ldots] \qquad (3.4\text{-}3)$$

according to the *Baker-Hausdorff formula* [Wiess and Maradudin, 1962]. For accuracy up to first order in Δz, we have

$$\exp[(\hat{D} + \hat{S})\Delta z] \simeq \exp(\hat{D}\Delta z)\exp(\hat{S}\Delta z) \qquad (3.4\text{-}4)$$

which implies that in Eq. (3.4-2) the diffraction and the inhomogeneous operators can be treated independent of each other and we can write Eq. (3.4-2) as

$$\psi_e(x, y; z + \Delta z) = \exp(\hat{S}\Delta z)\exp(\hat{D}\Delta z)\psi_e(x, y; z). \tag{3.4-5}$$

The action of the first operator on the right-hand side (RHS) of Eq. (3.4-5) is better understood in the spectral domain. Note that this is the propagation operator that takes into account the effect of diffraction between planes z and $z + \Delta z$. Propagation is readily handled in the spectral or spatial frequency domain using the transfer function for propagation given by Eq. (3.3-8) with z replaced by Δz. The second operator describes the effect of propagation in the absence of diffraction and in the presence of medium inhomogenieties, either intrinsic or induced and is incorporated in the spatial domain. Hence, the execution of the exponential operation $\exp(\hat{D}\Delta z)$ is carried out in the Fourier domain using the prescription

$$\exp(\hat{D}\Delta z)\psi_e = \mathcal{F}^{-1}\left\{ \exp\left[\frac{j(k_x^2 + k_y^2)\Delta z}{2k_0} \right] \mathcal{F}\{\psi_e\} \right\} \tag{3.4-6}$$

and the algorithm for a single step in Δz can be written as

$$\psi_e(x, y; z + \Delta z) = \exp(\hat{S}\Delta z)\exp(\hat{D}\Delta z)\psi_e(x, y; z)$$

$$= \exp(-j\Delta n k_0 \Delta z)\mathcal{F}^{-1}\left\{ \exp\left[\frac{j(k_x^2 + k_y^2)\Delta z}{2k_0} \right] \mathcal{F}\{\psi_e(x, y; z)\} \right\}$$

$$\tag{3.4-7}$$

The BPM repeats the above process until the field has traveled the desired distance. A schematic flow diagram of the BPM in its simplest form is shown in Fig. 3.2, which shows a recursive loop that is iterated until the final distance is warranted.

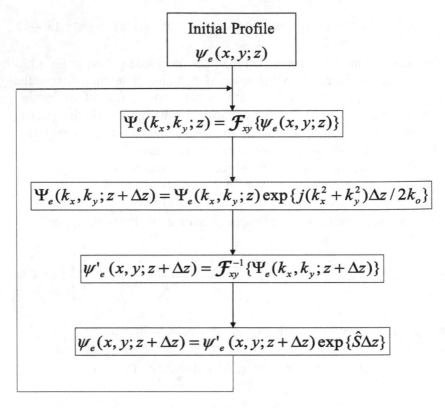

Fig. 3.2 Flow diagram for the beam propagation method.

Figure 3.3 offers an alternative illustration of the BPM. The upper part of the figure schematically shows the optical path is broken into a series of finite steps, where the down-arrows and up-arrows represent the Fourier transformation and inverse transformation, respectively. The straight arrows represent diffraction for a distance of Δz. The lower part of the figure, however, illustrates a physical process to represent BPM in that a plane wave is incident upon a transparency $t(x, y) = \psi_e(x, y; z = 0) = \psi_{e0}(x, y)$. The diffracted field after a distance of Δz is phase-modulated by a phase screen, $\exp[-j\Delta n(x, y)k_0\Delta z]$, and the resulting field is diffracted another distance of Δz before being phase-modulated again and then diffracted, and so on as shown in the figure until the desired distance is traveled to obtain the final output,

$\psi_e(x, y; z = m\Delta z)$, where m is some integer. This physical process clearly illustrates that the processes of diffraction and phase modulation are independent of each other.

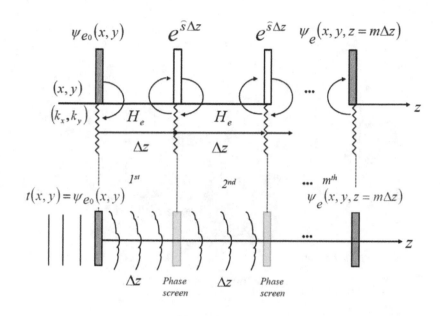

Fig. 3.3 Alternative illustration and physical interpretation of the BPM.

3.5 MATLAB Examples Using the Split-Step Beam Propagation Method

In this section, we demonstrate two examples using the beam propagation method. In the first one, we show the focusing of a Gaussian beam with initially plane phase-fronts of the form given by Eq. (3-2-23) by a thin lens with its phase transformation function, $\exp[j\frac{k_0}{2f}(x^2 + y^2)]$, as previously given by Eq. (2.3-23). We basically first have the Gaussian beam multiplied by the phase function of the lens and then let the beam propagation using the beam propagation method with $\exp(\hat{S}\Delta z) = 1$ in the algorithm given by Fig. 3.2. Table 3.1 provides the m-file for this example. Figure 3.4(a) and (b) are the results generated from Table 3.1. They show the focusing of the Gaussian beam of wavelength $\lambda_0 = 0.633\mu m$ with waist $w_0 = 10\,mm$ by the lens of focal length $f = 1600\,cm$. The lens is located at $z = 0$. The figures clearly show the focusing at $z = 1600\,cm$.

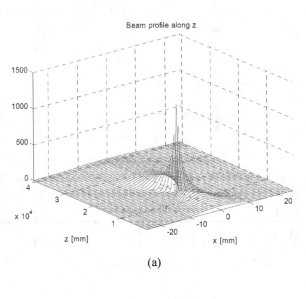

(a)

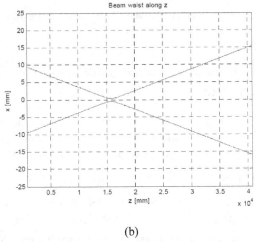

(b)

Fig. 3.4. (a) Showing beam profile along z and (b) Showing the beam waist along z, the waist is initially at 10 mm ($z = 0$) just before the thin lens of focal length of 1600 cm.

Table 3.1 BPM_focusing_lens.m (m-file for the simulation of Gaussian beam focused by a lens Using BPM).

```
%BPM_focusing_lens.m
%Simulation of Gaussian Beam Focused by a Lens Using BPM
%Paramters suggested for simulation :
% Ld (light wavelength) =0.633, wo (waist)=10,
% dz(sample distance along z)=800, Z(total final distance away from lens) =40000,
% f(focal length)=16000
clear

%Gaussian Beam
            N=255; %N : sampling number
            L=50*10^-3; %Display area
            Ld=input('wavelength of light in [micrometers] = ?');
            Ld=Ld*10^-6;
            ko=(2*pi)/Ld;
            wo=input('Waist of Gaussian Beam in [mm] = ?');
            wo=wo*10^-3;
            dz=input('step size of z (dz) in [mm] = ?');
            dz=dz*10^-3;
            Z=input('destination of z in [mm] = ? ');
            Z=Z*10^-3;
%Focal length of Lens
f=input('Focal length of lens in [mm]= ?');
f=f*10^-3;
% dx : step size
            dx=L/N;
            for n=1:256
                        for m=1:256
            %Space axis
            x(m)=(m-1)*dx-L/2;
            y(n)=(n-1)*dx-L/2;
%Frequency axis
            Kx(m)=(2*pi*(m-1))/(N*dx)-((2*pi*(256-1))/(N*dx))/2;
            Ky(n)=(2*pi*(n-1))/(N*dx)-((2*pi*(256-1))/(N*dx))/2;
                        end
            end

[X,Y]=meshgrid(x,y);
[KX, KY]=meshgrid(Kx,Kx);

%Gaussian Beam in space domain
Gau_ini=(1/(wo*pi^0.5))*exp(-(X.^2+Y.^2)./(wo^2));

%Energy of the initial Gaussian beam
```

```
Energy_ini=dx*dx*sum(sum(abs(Gau_ini).^2))

%Lens Equation
L=exp(j*ko/(2*f)*(X.^2+Y.^2));

%Gaussian Beam passed through the lens
Gau_ini=Gau_ini.*L;

%Free space transfer function of propagation
H=exp(j/(2*ko)*dz*(KX.^2+KY.^2));

%Iterative Loop
Gau=Gau_ini;
n1=0;


for z=0:dz:Z
            n1=n1+1;
            Zp(n1)=z+dz;
            %Gaussian Beam in Frequency domain
               FGau=fft2(Gau);
               %Propagated Gaussian beam in Frequency domain
               FGau=FGau.*fftshift(H);
               %Propagated Gaussian beam in space domain
               Gau=ifft2(FGau);
            %Step propagation through medium
               Gau_pro(:,n1)=Gau(:,127);
      end

%Energy of the final propagated Gaussian beam,
%to check conservation of energy
Energy_pro=dx*dx*sum(sum(abs(Gau).^2))

%axis in mm scale
x=x*10^3;
y=y*10^3;
Zp=Zp*10^3;
MAX1=max(max(abs(Gau_ini)));
MAX2=max(max(abs(Gau)));
MAX=max([MAX1 MAX2]);

figure(1);
mesh(x,y,abs(Gau_ini))
title('Initial Gaussian Beam')
xlabel('x [mm]')
ylabel('y [mm]')
axis([min(x) max(x) min(y) max(y) 0 MAX])
```

```
axis square

figure(2);
mesh(x,y,abs(Gau))
title('Propagated Gaussian Beam at Z')
xlabel('x [mm]')
ylabel('y [mm]')
axis([min(x) max(x) min(y) max(y) 0 MAX])
axis square

figure(3)
for l=1:n1
    plot3(x',Zp(l)*ones(size(x')),abs(Gau_pro(:,l)))
    hold on
end
  axis([min(x) max(x) min(Zp) max(Zp)])
grid on
title('Beam profile along z')
xlabel('x [mm]')
ylabel('z [mm]')
hold off

figure(4)
A=max(abs(Gau_pro));
B=diag(1./A);
N_Gau_pro=abs(Gau_pro)*B;
contour(Zp,x,N_Gau_pro, [ exp(-1) exp(-1)], 'k')
grid on
title('Beam waist along z')
xlabel('z [mm]')
ylabel('x [mm]')
```

In the second example, we simulate the propagation of a Gaussian beam through a square-law medium given by

$$n^2(x, y) = n_0^2 - n_2(x^2 + y^2) \tag{3.5-1}$$

using the BPM. We basically want to solve Eq. (3.4-1). We know diffraction is handled by $\hat{D} = \frac{1}{2jk_0}\nabla_t^2$, and $\hat{S} = -j\Delta nk_0$ is the inhomogeneous operator that needs to be found for the given medium. For the square-law medium, we can approximate

$$n(x,y) \approx n_0 - \frac{n_2}{2n_0}(x^2 + y^2) = n_0[1 + \Delta n(x,y)]$$

and hence,

$$\Delta n(x,y) = -\frac{n_2}{2n_0^2}(x^2 + y^2),$$

or the operator $\hat{S}$ becomes

$$\hat{S} = j\frac{n_2}{2n_0^2}(x^2 + y^2)k_0 \qquad (3.5\text{-}2)$$

for the square-law medium. Figure 3.5 shows the simulation of a Gaussian beam propagation in the square-law medium. Table 3.2 shows the m-file for generating the figures. The simulation results illustrate some of the conclusions obtained in Section 3.2, where it concludes that the Gaussian beam propagates along the square-law medium with its waist modulating with a period $z_m = \pi h$ with h again defining the square-law medium given by Eq. (3.2-1). To verify the numerical plots, let us relate h to n_0 and n_2. Using Eqs. (3.2-1) and (3.5-1), we can find the following relation:

$$h = \frac{n_0}{\sqrt{n_2}}. \qquad (3.5\text{-}3)$$

Hence the modulation period z_m becomes

$$z_m = \frac{\pi n_0}{\sqrt{n_2}}. \qquad (3.5\text{-}4)$$

For $n_0 = 1.5$ and $n_2 = 0.01\ m^{-1}$, $z_m = 47.123m$, which is consistent with the results given by the plots. From Eq. (3.2-14b), we can also find the waist of the fundamental mode

$$w_{00} = \sqrt{2}/\gamma = \sqrt{\frac{2n_0}{k_0(n_2)^{\frac{1}{2}}}} \qquad (3.5\text{-}5)$$

by using Eq. (3.2-5). For $\lambda_0 = 2\pi n_0/k_0 = 0.633\mu m$, $w_{00} \approx 1.419mm$. Note that for a Gaussian input waist of 5 mm, which is larger than w_{00}, we have the propagating Gaussian beam initially focused before defocused as shown in Figs. 3.5(a) and (b). For a Gaussian input waist of 1 mm, which is smaller than w_{00}, we have the propagating Gaussian beam initially defocused before focused, as clearly illustrated in Fig. 3.5(c). Indeed, this is the property of a square-law medium known as *periodic focusing*, which has been investigated using ray optics in Chapter 2 [see Problem 2.18].

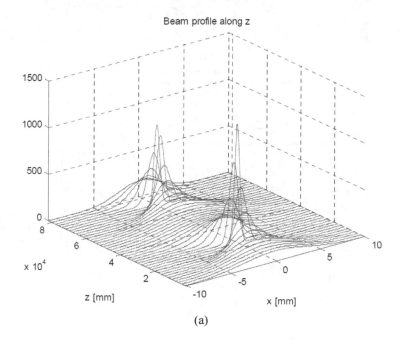

(a)

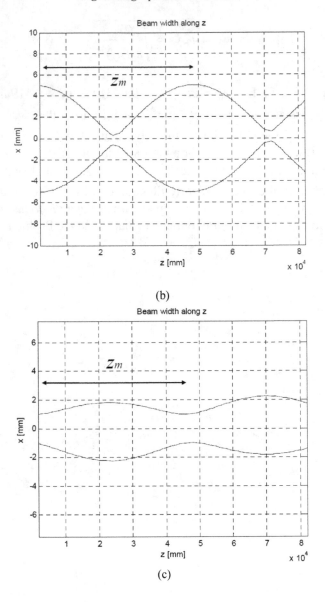

(b)

(c)

Fig. 3.5 Periodic focusing: (a) Showing Gaussian beam profile ($w_0 = 5\,\text{mm} > w_{00}$ at $z = 0$)
along z in a square-law medium ($n_0 = 1.5$, $n_2 = 0.01$), wavelength used is 0.633 μm,
and (b) Showing the beam waist along z, (c) Showing the beam waist when the input
Gaussian waist is 1 mm $< w_{00}$ at $z = 0$.

Table 3.2 BPM_sq_law_medium.m (m-file for the simulation of Gaussian beam propagating in a square-law medium using BPM).

```
%BPM_sq_law_medium.m
%Simulation of Gaussian Beam Propagation in a Square-Law Medium Using BPM
%This program demonstrates periodically focusing and defocusing
% Suggested simulation parameters
% For initially converging situation:
% L=20
% Ld=0.633;
% wo=5;
% dz=2000
% Z=80000
% For initially diverging situation:
% L=15
% Ld=0.633;
% wo=1;
% dz=2000

% Z=80000
clear
n0=1.5;
n2=0.01; % with unit of (meter)^-2
%Gaussian Beam
            %N : sampling number
            N=255;
            L=input('Length of back ground in [mm] = ?');
            L=L*10^-3;
            Ld=input('Incident wavelength of light in [micrometers] = ?');
            Ld=(Ld/n0)*10^-6;
            ko=(2*pi)/Ld; %wavenumber in n0
            wo=input('Waist of Gaussian Beam in [mm] = ?');
            wo=wo*10^-3;
            dz=input('step size of z (dz) in [mm] = ?');
            dz=dz*10^-3;
            Z=input('destination of z in [mm] = ? ');
            Z=Z*10^-3;
% dx : step size
            dx=L/N;

for n=1:256
   for m=1:256
            %Space axis
            x(m)=(m-1)*dx-L/2;
            y(n)=(n-1)*dx-L/2;
            %Frequency axis
            Kx(m)=(2*pi*(m-1))/(N*dx)-((2*pi*(256-1))/(N*dx))/2;
            Ky(n)=(2*pi*(n-1))/(N*dx)-((2*pi*(256-1))/(N*dx))/2;
```

```
    end
end

[X,Y]=meshgrid(x,y);
[KX, KY]=meshgrid(Kx,Kx);

%Gaussian Beam in space domain
Gau_ini=(1/(wo*pi^0.5))*exp(-(X.^2+Y.^2)./(wo^2));

%Energy of the initial Gaussian beam
Energy_ini=dx*dx*sum(sum(abs(Gau_ini).^2))

%Free space transfer function of step propagation
H=exp(j/(2*ko)*dz*(KX.^2+KY.^2));

%S operator according to index profile of medium
S=-j*(-n2/(2*n0*n0)*(X.^2+Y.^2))*ko;


    %Iterative Loop
    Gau=Gau_ini;
    n1=0;
    for z=0:dz:Z
                n1=n1+1;
                Zp(n1)=z+dz;
                %Gaussian Beam in Frequency domain
                FGau=fft2(Gau);
                %Propagated Gaussian beam in Frequency domain
                FGau=FGau.*fftshift(H);
                %Propagated Gaussian beam in space domain
                Gau=ifft2(FGau);
            %Step propagation through medium
            Gau=Gau.*exp(S.*dz);
            Gau_pro(:,n1)=Gau(:,127);
    end

%Energy of the final propagated Gaussian beam
Energy_pro=dx*dx*sum(sum(abs(Gau).^2))

%axis in mm scale
x=x*10^3;
y=y*10^3;
Zp=Zp*10^3;
MAX1=max(max(abs(Gau_ini)));
MAX2=max(max(abs(Gau)));
MAX=max([MAX1 MAX2]);

figure(1);
```

```
mesh(x,y,abs(Gau_ini))
title('Initial Gaussian Beam')
xlabel('x [mm]')
ylabel('y [mm]')
axis([min(x) max(x) min(y) max(y) 0 MAX])
axis square

figure(2);
mesh(x,y,abs(Gau))
title('Propagated Gaussian Beam')
xlabel('x [mm]')
ylabel('y [mm]')
axis([min(x) max(x) min(y) max(y) 0 MAX])
axis square

figure(3)
for l=1:n1

    plot3(x',Zp(l)*ones(size(x')),abs(Gau_pro(:,l)))
    hold on
end
grid on
axis([min(x) max(x) min(Zp) max(Zp)])
title('Beam profile along z')
xlabel('x [mm]')
ylabel('z [mm]')
hold off

w00=(2/((ko/n0)*(n2^0.5)))^0.5; %fundamental mode width
w00=w00*10^3;
sprintf('w00, fundamental mode width in [mm] %f' , w00)

figure(4)
A=max(abs(Gau_pro));
B=diag(1./A);
N_Gau_pro=abs(Gau_pro)*B;
contour(Zp,x,N_Gau_pro, [ exp(-1) exp(-1)], 'k')
grid on
title('Beam width along z')
xlabel('z [mm]')
ylabel('x [mm]')

zm=pi*n0/((n2^0.5)); %modulation period
zm=zm*10^3;
sprintf('zm, modulation period in [mm] %f' , zm)
```

3.6 Beam Propagation in Nonlinear Media: The Kerr Media

In Section 3.5, we have demonstrated that by solving Eq. (3.3-9) numerically, we can treat diffraction and the effect of inhomogeneity separately using the BPM. Indeed we see that the two independent equations,

$$\partial \psi_e / \partial z = \hat{D} \psi_e \qquad (3.6\text{-}1)$$

and

$$\partial \psi_e / \partial z = \hat{S} \psi_e, \qquad (3.6\text{-}2)$$

provide solutions to Fresnel diffraction and phase modulation independently. The solution to Eq. (3.6-2) corresponds to phase modulation as it is obvious that $\psi_e(x, y, z) = \exp(-j\Delta n k_0 z)$ is a solution. So in order to investigate the effect of beam propagation in the square-law medium, we can heuristically construct an equation by simply adding the operators on the right-hand side of Eqs. (3.6-1) and (3.6-2):

$$\partial \psi_e / \partial z = (\hat{D} + \hat{S}) \psi_e,$$

which has now become Eq. (3.4-1). By the same token, in order to investigate beam propagation in nonlinear media, we do the same by finding or modeling a nonlinear operator $\hat{N}$ such that

$$\partial \psi_e / \partial z = \hat{N} \psi_e \qquad (3.6\text{-}3)$$

represents the nonlinear effect on the propagating beam. The overall equation for propagation in a nonlinear medium is then simply found by constructing an equation such that the appropriate operators on the right hand side of Eq. (3.6-1) and (3.6-3) are added to give

$$\partial \psi_e / \partial z = (\hat{D} + \hat{N}) \psi_e. \qquad (3.6\text{-}4)$$

In this section, we discuss an important nonlinear effect called the *Kerr effect* and demonstrate the use of the BPM to solve this type of problem. The Kerr effect can be described by the following nonlinear dependence of the refractive index on the electric field of the light beam:

$$n = n_0 + n_{2E} |\psi_e|^2, \tag{3.6-5}$$

where n_0 is the refractive index of the medium in the absence of the light field, $n_{2E} |\psi_e|^2$ is the nonlinear variation of the index. n_{2E} is the *Kerr constant* and has a unit of $(m/V)^2$. The medium that can produce such an effect is called the *Kerr medium*.

Similar to the derivation of $\hat{S}$ in Eq. (3.5.2), we can obtain

$$\hat{N} = -j\Delta n k_0 = -j \frac{n_{2E}}{n_0} k_0 |\psi_e|^2 \tag{3.6-6}$$

for the Kerr medium. Eq. (3.6-4) then becomes

$$\frac{\partial \psi_e}{\partial z} = \frac{1}{2jk_0} \nabla_t^2 \psi_e - j \frac{n_{2E}}{n_0} k_0 |\psi_e|^2 \psi_e. \tag{3.6-7}$$

This equation is called the *nonlinear Schrödinger equation*, which describes the evolution of complex envelope $\psi_e(x, y, z)$ of a beam propagating along the z-direction for a given initial envelope $\psi_e(x, y, 0)$.

3.6.1 *Spatial soliton*

We cannot solve Eq. (3.6-7) using the Fourier transform techniques that were used to solve Eq. (3.3-6) because of the nonlinear term on the right side of the equation. However, analytical solutions do exist in one transverse dimension, including a well-known stable solution called *spatial soliton*. Solitons have a special property that they can propagate without changing their shape. In the quest for the expression for $|\psi_e|$ that does not depend on z, we substitute

$$\psi_e(x, y, z) = a(x, y) \exp(-j\kappa z) \tag{3.6-8}$$

in Eq. (3.6-7) to get

$$\nabla_t^2 a = -2\kappa k_0 a - (2k_0^2\, n_{2E}/n_0)a^3. \tag{3.6-9}$$

Consider now that we have one transverse dimension, namely, x. In this case, Eq. (3.6-9) becomes

$$\frac{d^2 a}{dx^2} = -2\kappa k_0 a - (2k_0^2 n_{2E}/n_0)a^3. \tag{3.6-10}$$

Multiplying both sides by $2da/dx$ and integrating, we have

$$\left(\frac{da}{dx}\right)^2 = -2\kappa k_0 a^2 - (2k_0^2 n_{2E}/n_0)a^4,$$

where we have neglected the integration constant. We can recast the above equation in the form

$$x = \int \frac{da}{\sqrt{-2\kappa k_0 a^2 - (k_0^2 n_{2E}/n_0)a^4}}, \tag{3.6-11}$$

which has a solution in the form

$$a(x) = A\sec h(Kx), \tag{3.6-12}$$

where

$$A = \sqrt{\frac{-2\kappa n_0}{k_0 n_{2E}}} \quad \text{and} \quad K = \sqrt{\frac{1}{-2\kappa k_0}}.$$

Equation (3.6-12) is the well-known *soliton* solution. A normalized plot of the sech-beam is shown in Fig. 3.6.

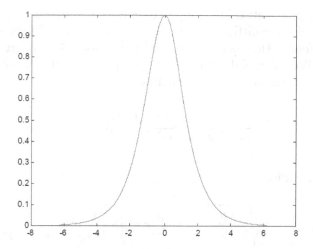

Fig. 3.6 Plot of sech(x), showing the shape of a spatial soliton in one transverse dimension.

We note from Eq. (3.6-12) that, $\kappa < 0$ and $n_{2E} > 0$ for a physical solution. This makes sense since for $n_{2E} > 0$, the refractive index is greater where the beam amplitude is higher [see Eq. (3.6-5)]. Note that the sech-beam is not a function of z, and hence non-diffracting. For such a beam shape, this means that the center of the beam sees the highest refractive index, which then decays off-axis. Explaining it in terms of ray optics, the light rays from the edge of the beam (lower index) tend to bend towards the center of the beam (higher index) through the Kerr effect [due to the second term of the right-hand side of Eq. (3.6-7)] and at the same time the beam itself undergoes diffraction [due to the first term of the right-hand side of Eq. (3.6-7)]. Thus, the nonlinearly induced inhomogeniety causes the beam to maintain its shape by exactly balancing the effects of spreading due to diffraction, and hence the shape of the sech-beam remains the same as it propagates inside the medium. The sech-beam belongs in a family commonly known as *non-spreading* or *diffraction-free* beams in a cubically nonlinear medium.

Limiting to one transverse dimension and for $n_{2E} > 0$, calculations also have shown periodic behavior (*periodic self-focusing*) if focusing due to the nonlinear effect is slightly larger than the diffraction effect. For $n_{2E} < 0$, we have *self-defocusing* [Korpel *et al.* (1986)]. At the end of this Chapter, we present some simulation results.

For propagation involving two transverse directions, i.e., x and y, the problem is more difficult and it appears that no closed-form solutions have been found. Here we only consider the case where there is radial symmetry. We express the transverse Laplacian in polar coordinates as [see Eq. (2.2-8) for the Laplacian operator]

$$\nabla_t^2 = \frac{\partial^2}{\partial x^2} + \frac{\partial^2}{\partial y^2} = \frac{d^2}{dr^2} + \frac{1}{r}\frac{d}{dr}. \tag{3.6-13}$$

Using the definitions

$$a = \sqrt{\frac{-\kappa n_0}{2k_0 n_{2E}}}\,\hat{a} \quad \text{and} \quad r = \sqrt{-\frac{1}{2\kappa k_0}}\,\hat{r}, \tag{3.6-14}$$

we can rewrite (3.6-9) in the form

$$\frac{d^2\hat{a}}{d\hat{r}^2} + \frac{1}{\hat{r}}\frac{d\hat{a}}{d\hat{r}} - \hat{a} + \hat{a}^3 = 0. \tag{3.6-15}$$

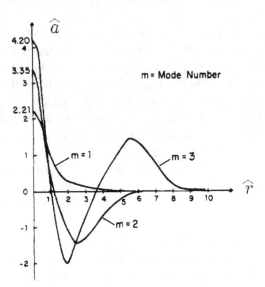

Fig. 3.7 Numerical solutions of Eq. (3.6-15) showing different modes [Adopted from Korpel and Banerjee (1984)].

Again, this nonlinear ordinary differential equation has no analytical solutions; the solutions are obtained by numerical methods with boundary conditions $\hat{a}(\infty) \to 0$ and $d\hat{a}(0)/d\hat{r} = 0$ [Chiao, Garmire and Townes (1964), Haus (1966), Korpel and Banerjee (1984)]. These are shown in Fig. 3.7. These solutions give rise to the so-called *self-trapping* phenomenon. Note that the solutions are multimodal in nature, with the mode number m depending on the initial condition $\hat{a}(0)$. In a way, these solutions are reminiscent of the modes in the square-law medium, discussed in Section 3.2. Note that for the first mode, i.e., m=1, $\hat{a}(0) = 2.21$ and the mode resembles the shape of a sech-beam in one transverse dimension.

3.6.2 Self-focusing and self-defocusing

As pointed out in the last Section, through the Kerr effect, light rays from the edge of the beam tend to bend toward to the center of the beam and this nonlinear optical phenomenon is called *self-focusing* (and in some cases defocusing) [Kelly (1965)]. The self-focusing dynamics of high-power ultrashort laser pulses has attracted significant attention in recent years, particularly for the case in which ultrashort pulses propagate through the atmosphere [Tzortzakis *et al.* (2001)] or for the case of ultrafast laser beam delivery for materials processing [Sun and Longtin (2004)]. Indeed self-focusing can cause optical damage even at lower incident intensities than might otherwise be expected. In this Section, we estimate the necessary power required for self-focusing and provide an estimate of the focal length.

Let us assume the incident laser beam has a Gaussian distribution

$$\psi_e(x,y) = \psi_0 \exp\left[-(x^2 + y^2)/w_0^2\right].$$

Hence, for $n_{2E} > 0$, we have, from Eq. (3.6-5)

$$n(x,y) = n_0 + n_{2E}\psi_0^2 \exp[-2(x^2 + y^2)/w_0^2], \qquad (3.6\text{-}16)$$

which clearly means that intensity distribution creates a refractive index profile within the beam having a maximum value on the center and gradually decreasing away from the center. Expanding the exponent and keeping the first two terms, Eq. (3.6-16) becomes

$$n(x,y) = (n_0 + n_{2E}\psi_0^2) - n_{2E}\psi_0^2 2(x^2 + y^2)/w_0^2$$

$$\approx n_0 - \frac{2n_{2E}\psi_0^2}{w_0^2}(x^2 + y^2)$$

as $n_{2E}\psi_0^2 \ll n_0$, which is true in practice. Now we want to square $n(x,y)$ such that

$$n^2(x,y) \approx n_0^2 - \frac{4n_0 n_{2E}\psi_0^2}{w_0^2}(x^2 + y^2) \qquad (3.6\text{-}17)$$

induced by the Kerr effect can be compared with the square-law medium given below by

$$n^2(x,y) \approx n_0^2 - n_2(x^2 + y^2). \qquad (3\text{-}6\text{-}18)$$

We can see immediately that the effect of the Kerr medium can be modeled by a square-law medium, at least under the paraxial approximation. We have pointed out that the square-law medium has periodic focusing property with modulation period z_m given by Eq. (3.5-4), which is

$$z_m = \frac{\pi n_0}{\sqrt{n_2}}.$$

$$(3.6\text{-}19)$$

But now we have $n_2 = 4n_0 n_{2E}\psi_0^2 / w_0^2$ by comparing Eqs. (3.6-17) and (3.6-18). Hence we can predict the Kerr medium will have a focusing length given approximately by

$$f_{NL} = \frac{z_m}{2} \approx \frac{\pi n_0}{2\sqrt{n_2}} = \frac{\pi}{4}\sqrt{\frac{n_0}{n_{2E}}}\frac{w_0}{\psi_0}. \qquad (3.6\text{-}20)$$

The subscript NL signifies that the focusing effect is due to nonlinear phenomenon. Therefore, we can consider the focusing of the Gaussian beam produces convergent rays with convergence angle given by

$$\theta_{NL} \approx \frac{w_0}{f_{NL}} = \frac{4}{\pi}\sqrt{\frac{n_{2E}}{n_0}}\psi_0. \qquad (3.6\text{-}21)$$

Now in the absence of any nonlinear effects, the optical beam will spread because of diffraction and the spread of the angle of a Gaussian beam has been calculated and given by Eq. (2.4-7) and the angle of spread in medium with index n_0 is then given by

$$\theta_{sp} = \frac{\lambda_0}{\pi w_0 n_0}. \tag{3.6-22}$$

We, therefore, expect that self-focusing can compete with diffraction. In fact, for $\theta_{NL} = \theta_{SP}$, the effects of self-focusing and diffraction will cancel each other and the beam will propagate without any focusing or defocusing. This phenomenon is known as *self-trapping* as pointed out previously. Indeed, the first mode ($m=1$) shown in Fig. 3.7 is a self-trapped beam first calculated by Chaio, Garmire and Townes (1964).

We shall now estimate the power required for self-trapping. Such a power is termed the *critical power P_{cr}*. From Eq. (2.2-34), which is the equation of power density for a plane wave with amplitude E_0, we can immediately write down the power density for a Gaussian beam propagating along the z-direction as

$$<\mathbf{S}> = \frac{|\psi_0|^2}{2\eta} \exp[-2(x^2 + y^2)/w_0^2]\mathbf{a_z}$$

$$= \frac{|\psi_0|^2}{2\eta} \exp(-2r^2/w_0^2)\mathbf{a_z},$$

where η is the characteristic impedance of the medium of index n_0. The total power of the beam is then given by

$$P_0 = \int_0^\infty <\mathbf{S}> \cdot 2\pi r dr \, \mathbf{a_z}$$

$$= \frac{n_0 \varepsilon_0 c \pi |\psi_0|^2 w_0^2}{4}. \tag{3.6-23}$$

To find the critical power, we calculate the required field amplitude by equating $\theta_{NL} = \theta_{sp}$ to get

$$\psi_{0,cr} = \frac{\lambda_0}{4w_0\sqrt{n_0 n_{2E}}}. \tag{3.6-24}$$

By substituting Eq. (3.6-24) into Eq. (3.6-23), we have the critical power

$$P_{cr} = \frac{\pi c \varepsilon_0 \lambda_0^2}{64 n_{2E}}. \tag{3.6-25}$$

Hence for $P_0 = P_{cr}$, we have self-trapping. For $P_0 < P_{cr}$, diffraction will dominate and the beam will diverge. For $P_0 > P_{cr}$, the nonlinear effects will dominate and the beam will self-focus, which may tend to become anomalously large in the self-focused region and cause breakdown of the medium. However, in some cases, the high intensity of the laser beam could ionize the medium and create electron plasma that tends to counteract self-focusing, stabilizing the intense region into a *filament* or filaments [Braun *et al.* (1995), Tzortzakis *et al.* (2001)]. To investigate such phenomenon, one could extend the nonlinear Schrödinger equation by simply adding relevant terms to include other effects. Other effects could include not only the plasma effect [Sun and Longtin (2004), Fibich *et al.* (2004)], but also higher order nonlinearity as well as two-photon absorption [Centurion, Pu and Psaltis (2005)]. However, these effects are beyond the scope of this book.

　　Let us now return to Eq. (3.6-25). We note that the unit of n_{2E} is $(m/V)^2$ since, according to Eq. (3.6-5), ψ_e has units of V/m. However, in the modern literature, Eq. (3.6-5) is equivalently written as

$$n = n_0 + n_{2E}|\psi_e|^2 = n_0 + n_{2I}I, \tag{3.6-26}$$

where I is the intensity of the optical beam, which has units of W/m^2 or W/cm^2. In order to make the connection between n_{2E} and n_{2I}, we recognize that, from Eq. (2.2-36), $I = \varepsilon u |\psi_e|^2 / 2 = n_0 c \varepsilon_0 |\psi_e|^2 / 2$, and hence

$$n_{2E} = \frac{1}{2} n_0 c \varepsilon_0 n_{2I}.$$ (3.6-27)

Therefore, we can write Eq. (3.6-25) as

$$P_{cr} = \frac{\pi \lambda_0^2}{32 n_0 n_{2I}}.$$ (3.6-28)

Table 3.3 lists some of the physical properties of commonly used materials for the investigation of the Kerr effect.

Table 3.3 Physical properties of some materials.

Material	$n_{2E} \ (m/V)^2$	$n_{2I} (cm^2/W)$
Carbon disuphide (CS$_2$)	$5.8 \times 10^{-20} \ (n_0 = 1.6)$	3×10^{-15} (a)
Air (1 atm)	$6.1 \times 10^{-24} \ (n_0 = 1)$	5×10^{-19} (b)
Helium (1 atm)	$4.3 \times 10^{-26} (n_0 = 1)$	3.5×10^{-21}(b)

(a) Ganeer *et al.* Appl. Phys. B, 78, 433-438 (2004)
(b) Lide ed., CRC Handbook of Chemistry and Physics, 75th ed. (1994)

As an example, according to Eq. (3.6-28) and the use of Table 3.3, we can estimate the critical power for liquid CS_2 to be around 98KW for ruby-laser with wavelength of 0.6943 μm. Note also that according to Eq. (3.6-28), radiation of longer wavelength, such as radio waves, makes the critical power much higher and probably unattainable.

To end this chapter, we demonstrate some simulations using BPM for a 1-D Gaussian beam propagating in the Kerr medium. Table 3.4 shows the m-file for generating the figures. In the program, we have used a Gaussian beam of waist equal to 1 μm, and if we choose A=1, the amplitude of the Gaussian beam, which corresponds to $P_0 > P_{cr}$ and periodic self-focusing occurs. Figures 3.8(a)-(c) show such phenomenon from MATLAB simulations. In Fig. 3.8(d)-(f), we have chosen A=0.3, which corresponds to $P_0 < P_{cr}$ and it is clear that diffraction is dominant and no self-focusing is observed. Note that in this m-file, we simply try to demonstrate the Kerr effect using BPM and the parameters used do not necessarily correspond to the actual physical parameters and therefore all scales in the m-file are arbitrary.

Beam profile along z

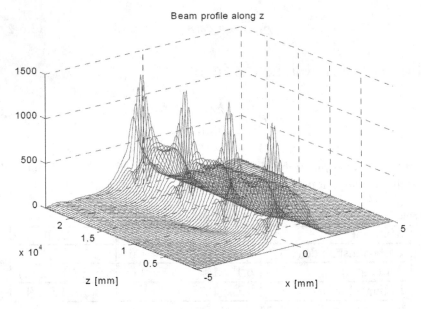

(a)

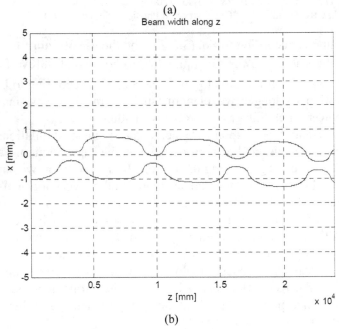

(b)

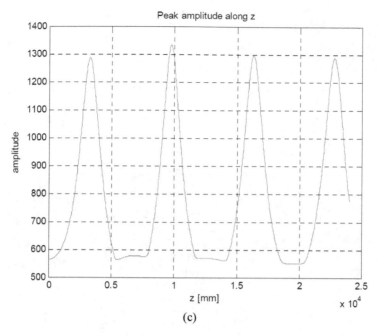

(c)

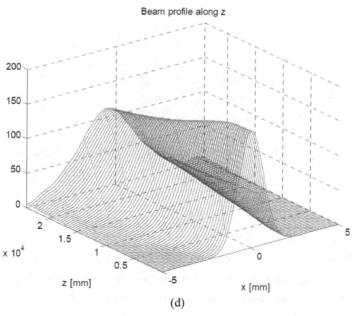

(d)

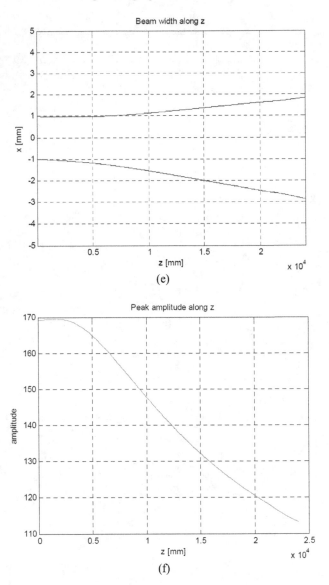

Fig. 3.8 Periodic focusing ($P_0 > P_{cr}$): (a) beam profile along z, (b) beam waist along z, (c) on-axis peak amplitude along z, and diffraction dominance ($P_0 < P_{cr}$), (d) beam profile along z, (e) beam waist along z; (f) on-axis peak amplitude along z.

Table 3.4 BPM_perioic_focusing.m (m-file for the simulation of a Gaussian beam propagating in the Kerr medium using BPM).

```
%BPM_Kerr effect.m
%Suggested simulation parameters
% wo=1;
% dz=25
% Z=24000
clear
D=0.6;
n0=1.5;
n2=2*10^-13; %proportional to Kerr constant
%Gaussian Beam
%N : sampling number
N=255;
L=10; %length of background in [mm]
L=L*10^-3;
Ld=0.633; %wavelength
Ld=(Ld/n0)*10^-6;
ko=(2*pi)/Ld; %wavenumber in n0
wo=input('Waist of Beam in [mm] = ?');
wo=wo*10^-3;
dz=input('step size of z (dz) in [mm] = ?');
dz=dz*10^-3;
Z=input('destination of z in [mm] = ? ');
Z*10^-3;
% dx : step size
dx=L/N;
    for m=1:256
      %Space axis
          x(m)=(m-1)*dx-L/2;

      %Frequency axis
          Kx(m)=(2*pi*(m-1))/(N*dx)-((2*pi*(256-1))/(N*dx))/2;
    End


%Gaussian Beam in space domain
A=input('Choose Amplitude of Gaussian Beam: 1 for self-periodic focusing, 0.3 for
diffraction dominance=');
  Gau_ini=A*(1/(wo*pi^0.5))*exp(-(x.^2)./(wo^2));
    %Energy of the initial Gaussian beam
    Energy_ini=dx*sum(abs(Gau_ini).^2)
    %Free space transfer function of step propagation
    H=exp(j/(2*ko)*dz*(Kx.^2));
    %Iterative Loop
    Gau=Gau_ini;
    n1=0;
    for z=0:dz:Z
```

```
    n1=n1+1;
    Zp(n1)=z+dz;
%Gaussian Beam in Frequency domain
            FGau=fft(Gau);
            %Propagated Gaussian beam in Frequency domain
            FGau=FGau.*fftshift(H);
            %Propagated Gaussian beam in space domain
            Gau=ifft(FGau);
%S operator according to index profile of medium
        S=-j*n2*ko*abs(Gau).^2;
        %Step propagation through medium
        Gau=Gau.*exp(S.*dz);
        Gau_pro(:,n1)=Gau';
end

%Energy of the final propagated Gaussian beam
Energy_pro=dx*sum(abs(Gau).^2)

%axis in mm scale


x=x*10^3;
Zp=Zp*10^3;
MAX1=max(abs(Gau_ini));
MAX2=max(abs(Gau));
MAX=max([MAX1 MAX2]);

figure(1);
plot(x,abs(Gau_ini))
title('Initial Beam')
xlabel('x [mm]')
ylabel('Amplitude')
axis([min(x) max(x) 0 MAX])
axis square
grid on

figure(2);
plot(x,abs(Gau))
title('Propagated Beam')
xlabel('x [mm]')
ylabel('amplitude')
axis([min(x) max(x) 0 MAX])
axis square
grid on

figure(3)
for l=1:15:n1
plot3(x',Zp(l)*ones(size(x')),abs(Gau_pro(:,l)))
```

```
        hold on
end
grid on
axis([min(x) max(x) min(Zp) max(Zp)])
title('Beam profile along z')
xlabel('x [mm]')
ylabel('z [mm]')
hold off

figure(4)
A=max(abs(Gau_pro));
B=diag(1./A);
N_Gau_pro=abs(Gau_pro)*B;
contour(Zp,x,N_Gau_pro, [ exp(-1) exp(-1)], 'k')
grid on
title('Beam width along z')
xlabel('z [mm]')
ylabel('x [mm]')

figure(5)


plot(Zp, max(abs(Gau_pro)))
grid on
title('Peak amplitude along z')
xlabel('z [mm]')
ylabel('amplitude')
```

Problems:

3.1   Verify Eq. (3.2-19) by substituting Eqs. (3.2-11) and (3.2-18) into the definition of $K(x, y, x', y')$ given by Eq. (3-2-17).

3.2   Show that, in a square-law medium, when $w_0 = w_{00} = \sqrt{2}/\gamma$, i.e., the incident Gaussian beam width is w_{00}, it excites only the fundamental mode with $w(z) = w_0$ which is independent of z. The beam propagates with its beam width unchanged.

3.3   Verify Eq. (3.2-24) by substituting Eqs. (3.2-23) and (3.2-19) into Eq. (3.2-17).

3.4   Show that when the Gaussian of waist w_0 propagates in a square-law medium, its width oscillates between w_0 and $2/w_0\gamma^2$ Thus to begin the propagation at z = 0, if $w_0 > w_{00}$, i.e., the incident beam waist is greater than the fundamental mode width, the beam shrinks from $w_0 (> w_{00})$ to $2/w_0\gamma^2 (< w_{00})$ and then back to w_0. We have periodic focusing and defocusing and this corresponds to the result obtained from ray optics mentioned in Problem 2.18.

3.5   When $h \to \infty$ in Eq. (3.2-1), the square-law medium corresponds to a homogeneous medium of refractive index n_0. Show that the waist of the Gaussian beam in Eq. (3.2-24) becomes

$$w^2(z) = w_0^2 + \left(\frac{4}{w_0^2 k_0^2} \right) z^2,$$

which represents the spreading of a Gaussian beam as it propagates through a homogeneous medium of refractive index n_0 as it is shown in Eq. (2.4-6b).

3.6 Solve Eq. (3.6-11) and show that Eq. (3.6-12) is a solution. To show it, use the substitution $y = \text{sech}\theta$ to simplify the integration of the form

$$\int \frac{dy}{y\sqrt{1-y^2}}.$$

3.7 Starting from the paraxial wave equation in the presence of a cubic nonlinearity, i.e., using Eq. (3.6-7) but limited to x and z dimensions, show that during the propagation of a complex envelope ψ_e, there is conservation of power. To do this, assume $\psi_e(x,z) = a(x,z)\exp[-j\phi(x,z)]$ and show that

$$\int_{-\infty}^{\infty} a^2(x,z) = \text{constant}.$$

3.8 Find the critical power for self-focusing in air at one atmospheric pressure with the use of laser with wavelength of 0.8 μm.

Bibliography

Agrawal, G. P. (1989). *Nonlinear Fiber Optics*. Academic Press, New York.

Banerjee, A. (2004). *Nonlinear Optics: Theory, Numerical Modeling, and Applications*, Marcel Dekker, Inc., New York.

Banerjee, P.P. and T.-C. Poon (1991). *Principles of Applied Optics.* Irwin, Illinois.

Braun, A., G. Korn, X. Liu, D. Du, J. Squier, and G. Mourou (1995). "Self-Channeling of High-Peak-Power Femtosecond Laser Pulses in Air," *Opt. Lett.* 20, pp. 73-75.

Centurion M., Y. Pu and D. Psaltis (2005). "Self-organization of Spatial Solitons," *Optics Express*, 12, pp. 6202-6211.

Chaio R.Y., E. Garmire, and C. H. Townes (1964). "Self-Trapping of Optical Beams," *Phys. Rev. Lett.*, 13, pp. 479-482.

Fibich G., S. Eisenmann, B. Ilan, and A. Zigler (2004). "Control of Multiple Filamentation in Air, " *Opt. Lett.*, 29, pp. 1772-1774.

Ghatak, A.K. and K. Thyagarajan (1989). *Optical Electronics*, Cambrigde University Press, Cambridge.

Haus., H. A., (1966). "Higher Order Trapped Light Beam Solutions," *Appl.Phys. Lett.*, 8, pp. 128-129.

Haus, H.A. (1984). *Waves and Fields in Optoelectronics.* Prentice-Hall, Inc.New Jersey.

Kelly, P.L. (1965). "Self-Focusing of Optical Beams," *Phys. Rev. Lett.*, 15, pp.1005-1008.

Korpel, A. and P.P. Banerjee (1984). "A Heuristic Guide to Nonlinear Dispersive Wave Equations and Soliton-Type Solutions, " *Proc. IEEE*, 72, pp.1109-1130.

Korepl, A., K.E. Lonngren, P.P. Banerjee, H.K. Sim, and M. R. Chatterjee (1986). "Split-Step-Type Angular Plane-Wave Spectrum Method for the study of Self-Refractive Effects in Nonlinear Wave Propagation," *J. Opt. Soc. Am. B*, 3, pp. 885-890.

Marcuse, D. (1982). *Light Transmission Optics.* Van Nostrand Reinhold Company, New York.

Poon T.-C. and P. P. Banerjee (2001). *Contemporary Optical Image Processing with MATLAB®.* Elsevier, Oxford, UK.

Sodha, M.S. and A.K. Ghatak (1977). *Inhomogeneous Optical Waveguides*. Plenum Press, New York.

Schiff, L. I. (1968). *Quantum Mechanics.* McGraw-Hill, New York.

Sun, J. and J.P. Longtin (2004). "Effects of a Gas Medium on Ultrafast Laser Beam Delivery and Materials Processing," *J. Opt. Soc. Am. B*, 21, pp. 1081-1088.

Tzortzakis, L. B., *et al*. (2001). "Breakup and Fusion of Self-Guided Femtosecond Light Pulses in Air," *Phys. Rev. Lett.*, 86, pp. 5470-5473.

Weiss, G. H. and A.A. Maradudin (1962). "The Baker-Hausdorff Formula and a Problem in Crystal Physics," *Journal of Mathematical Physics*, 3, pp. 771-777.

Chapter 4

Acousto-Optics

Acousto-optics deals with the interaction between sound and light. As a result of acousto-optic interaction, light waves can be modulated by electrical signals, which provides a powerful means for optically processing information. In fact, some of the key components in modern optical processors and optical display systems usually consist of one or more acousto-optic modulators. Acousto-optic interactions can result in laser beam deflection, laser intensity modulation, phase modulation as well as frequency shifting of laser.

4.1 Qualitative Description and Heuristic Background

An *acousto-optic modulator* (AOM) is a spatial light modulator that comprises an acoustic medium (such as glass or water) to which a piezoelectric transducer is bonded. Through the action of the piezoelectric transducer, the electrical signal is converted into sound waves propagating in the acoustic medium with an acoustic frequency spectrum that matches, within the bandwidth limitations of the transducer, that of the electrical excitation. The pressure in the sound wave creates a traveling wave of rarefaction and compression, which in turn causes perturbations of the index of refraction. Thus, the acousto-optic device shown in Fig. 4.1 may be thought to act as a thin (phase) grating with an effective grating line separation equal to the wavelength Λ of the sound in the acoustic medium. We know that a phase grating splits incident light into various diffracted orders and we can show that the directions of the diffracted or scattered light inside the "*sound cell*" are governed by the *grating equation*,

188

$$\sin\phi_m = \sin\phi_{inc} + m\lambda_0/\Lambda, \quad m = 0,\pm1,\pm2,\ldots, \quad (4.1\text{-}1)$$

where ϕ_m is the angle of the m-th diffracted order light beam, ϕ_{inc} is the angle of incidence, and λ_0 is the wavelength of the light, all measured in

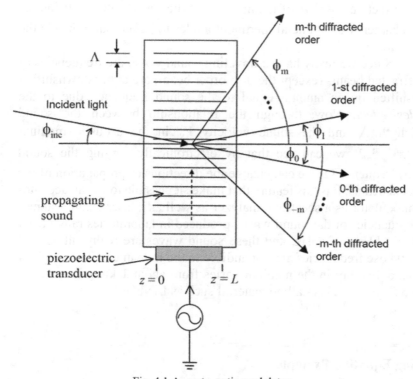

Fig. 4.1 Acousto-optic modulator.

the acoustic medium. The convention for angles is counterclockwise positive. Note that the angle between neighboring orders, as shown in Fig. 4.1, is twice the *Bragg angle*, ϕ_B:

$$\sin\phi_B = \frac{\lambda_0}{2\Lambda} = \frac{K}{2k_0}, \quad (4.1\text{-}2)$$

where $k_0 = |\boldsymbol{k_0}| = 2\pi / \lambda_0$ is the wavenumber of light inside the acoustic medium, and $K = |\mathbf{K}| = 2\pi / \Lambda$ is the wavenumber of sound. For example, we note that the angle measured between the zeroth order and the first order is $\sim 2\phi_B$, which is λ_0 / Λ as shown in Fig. 4.1. When measured outside the acoustic medium, these angles are increased through refraction by the refractive index n_0 of the material of the sound cell. In all figures in this chapter, we assume all pertinent angles to be measured inside the sound cell.

Since we really have a traveling sound wave, the frequencies of the diffracted beams (except the 0-th order beam) are either downshifted or upshifted by an amount equal to the sound frequency due to the *Doppler effect*. Now, through the relationship between the sound wavelength Λ and the sound velocity V_S in the acoustic medium $(\Lambda = 2\pi V_S / \Omega)$ we can see that by electronically varying the sound (radian) frequency Ω, we can change the directions of propagation of the diffracted beams. It is this feature that makes it possible to use an acousto-optic modulator as a spectrum analyzer as well as a laser beam scanner. The frequencies of the sound waves produced in laboratories range from about 100 kHz to 3 GHz, and these sound waves are really ultrasound waves whose frequencies are not audible to the human ear. The range of the sound velocity in the medium varies from about 1 km/s in water to about 6.5 km/s in a crystalline material such as $LiNbO_3$.

===

Grating Equation Example

A diffraction grating is composed of a large number of parallel slits, all with the same width and spaced equal distances d between centers. $N = 1/d$ is called the *grating constant* and it is the number of lines per unit length. A plane wave of wavelength λ_0 is incident on a diffraction grating at an angle ϕ_{inc}, as shown in the below figure (a). Let us find the conditions for maximum diffraction.

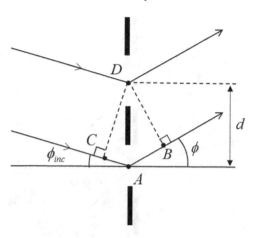

(a) Diffraction grating problem when the two sine terms in the path difference add.

For constructive interference, the path difference between the two diffracted parallel rays is $m\lambda_0$, where $m = 1, 2, \ldots$. So $CA + AB = m\lambda_0$, or

$$d\sin\phi_{inc} + d\sin\phi = m\lambda_0,\tag{a}$$

which can be written as

$$\sin\phi_m = -\sin\phi_{inc} + m\lambda_0/d,\tag{b}$$

where we have replaced ϕ by ϕ_m to reflect the fact that the diffracted angles is a function of m. Let us now find the conditions for maximum diffraction for the situation shown in the below figure (b). Again, for constructive interference, the path difference between the two diffracted parallel rays is $DB - CA = m\lambda_0$, which is

$$d\sin\phi - d\sin\phi_{inc} = m\lambda_0,\tag{c}$$

or, after replacing ϕ by ϕ_m, we have

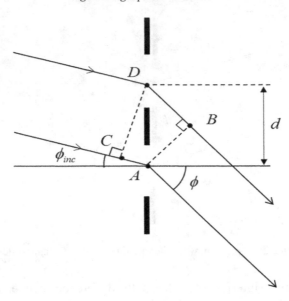

(b) Diffraction grating problem when the two sine terms in the path difference subtract.

$$\sin\phi_m = \sin\phi_{inc} + m\lambda_0\,/\,d\,. \tag{d}$$

From the preceding cases, we see that the two sine terms in the path difference may add or subtract [see Eqs. (a) and (c)], depending on the direction of the diffracted light. Hence the corresponding grating equations [Eqs. (b) and (d)] also carry different signs for different situations.

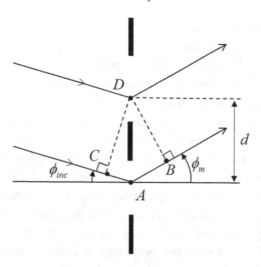

(c) Conventions for generalized grating equation given by Eq. (e). All the angles are measured from the horizontal axis.

However, the grating equation developed may be generalized for all angles of diffraction and incidence, but we need to adopt a sign convention for the angles. The generalized grating equation is given as follows:

$$\sin \phi_m = \sin \phi_{inc} + m\lambda_0 / d, \quad m = 0, \pm 1, \pm 2, ..., \tag{e}$$

All the angles are measured from the horizontal axis as shown in Fig. (c), and the convention for angles is counterclockwise positive. m is called the order of the diffraction maximum. For example, in the geometry in Fig. (b), $\phi_{inc} < 0$, $\phi_m < 0$, and m is negative. Considering -1st diffracted order, i.e., we take $m = -1$, Equation (e) becomes

$$\sin(-\phi_m) = \sin(-\phi_{inc}) - \lambda_0 / d,$$

which recovers Eq. (d) as $m = 1$ is used in the equation.

The phase-grating approach of acousto-optic interaction is somewhat of an oversimplification in that the approach does not predict the required angle for incident light in order to obtain efficient operations, nor does it explain why only one order is generated for a sufficiently wide transducer (i.e., L is large in Fig. 4.1). Another approach considers the interaction of sound and light as a collision of photons and phonons. For these particles to have well-defined momenta and energies, we must assume that we have interaction of monochromatic plane waves of light and sound, that is, we assume that the width L of the transducer is sufficiently wide in order to produce plane wave fronts at a single frequency. In the process of collision, two conservation laws have to be obeyed, namely, the *conservation of energy* and the *conservation of momentum*. If we denote the propagation vectors (also called *wave vectors*) of incident plane wave of light, diffracted or scattered plane wave of light, and sound plane wave in the acoustic medium by k_0, k_{+1}, and K, respectively, as shown in Fig. 4.2, we can write the condition for conservation of momentum as

$$\hbar k_{+1} = \hbar k_0 + \hbar K,$$

where $\hbar = h/2\pi$ and h is Planck's constant. Dividing the above equation by $\hbar$ leads to

$$k_{+1} = k_0 + K. \qquad (4.1\text{-}3a)$$

The corresponding conversation of energy takes the form (after division by $\hbar$)

$$\omega_{+1} = \omega_0 + \Omega, \qquad (4.1\text{-}3b)$$

where $\omega_0, \Omega,$ and ω_{+1} are the (radian) frequencies of the incident light, sound, and scattered light, respectively. The interaction described by Eq. (4.1-3) is called the *upshifted interaction*. Figure 4.2(a) shows the *wave-vector diagram*, and Fig. 4.2(b) describes the diffracted beam being upshifted in frequency. Because for all practical cases $|K| << |k_0|$, the magnitude of k_{+1} is essentially equal to k_0 and therefore the wave-vector momentum triangle shown in Fig. 4.2(a) is nearly isosceles.

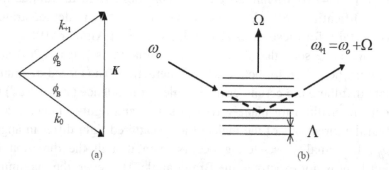

(a)   (b)

Fig. 4.2 Upshifted interaction: a) wave-vector diagram; b) experimental configuration.

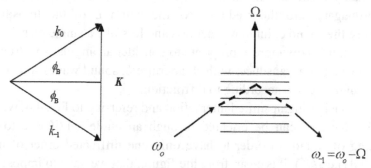

Fig. 4.3 Downshifted interaction: a) wave-vector diagram;
b) experimental configuration.

Suppose now that we change the directions of the incident and diffracted light as shown in Fig. 4.3. The conservation laws can be applied again to obtain two equations similar to Eqs. (4.1-3a) and (4.1-3b). The two equations describing the interaction are now

$$k_{-1} = k_0 - K, \tag{4.1-4a}$$

and

$$\omega_{-1} = \omega_0 - \Omega, \tag{4.1-4b}$$

where the subscripts on the left-hand side of the equation indicate that the interaction is *downshifted*.

There is some interesting physics hidden in Eqs. (4.1-3) and (4.1-4). It can be shown that Eqs. (4.1-3) refer to *phonon absorption* and Eqs. (4.1-4) to *stimulated phonon emission*. Indeed, attenuation and amplification of a sound wave have been demonstrated experimentally for these cases [Korpel, Adler, and Alpiner (1964)].

We have seen that the wave-vector diagrams [Figs. 4.2(a) and 4.3(a)] must be closed for both cases of interaction. The closed diagrams stipulate that there are certain critical angles of incidence ($\phi_{inc} = \pm \phi_B$) in the acoustic medium for plane waves of sound and light to interact, and also that the directions of the incident and scattered light differ in angle by $2\phi_B$. In actuality, scattering occurs even though the direction of incident light is not exactly at the Bragg angle. However, the maximum scattered intensity occurs at the Bragg angle. The reason is that we do not have exactly plane wavefronts; the sound waves actually spread out as they propagate into the medium. As the width L of the transducer decreases, the sound column will act less and less like a single plane wave and, in fact, it is now more appropriate to consider an angular spectrum of plane waves. For a transducer with an aperture L, sound waves spread out over an angle $\pm \Lambda / L$ according to diffraction.

Considering upshifted interaction and referring to Fig. 4.4, we see that the K-vector can be oriented through an angle $\pm \Lambda / L$ due to the spreading of sound. In order to have only one diffracted order of light generated (i.e., k_{+1}), it is clear, from the figure, that we have to impose the condition

$$\frac{\lambda_0}{\Lambda} >> \frac{\Lambda}{L},$$

or

$$L >> \Lambda^2 / \lambda_0. \tag{4.1-5}$$

This is because for k_{-1} to be generated, for example, a pertinent sound wave vector must lie along K'; however, this either is not present, or is present in negligible amounts in the angular spectrum of the sound (i.e., the downshifted wave vector triangle cannot be completed), if the condition, i.e., Eq. (4.1-5), is satisfied. If L satisfies Eq. (4.1-5), the acousto-optic modulator is said to operate in the *Bragg regime* and the device is commonly known as the *Bragg cell.*

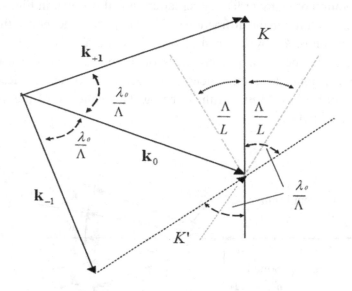

Fig.4.4 Wave-vector diagram illustrating the condition for defining the Bragg regime.

In the case where L is sufficiently short, we have the second regime of scattering (diffraction), called *Raman-Nath* (or *Debye-Sears*) *diffraction.* The condition

$$L \ll \Lambda^2 / \lambda_0, \qquad (4.1\text{-}6)$$

therefore, defines the *Raman-Nath regime.* In Raman-Nath diffraction, k_{+1} and k_{-1} (i.e., positive and negative first order scattered light) are generated simultaneously because various directions of plane waves of sound are provided from a small-aperture transducer.

So far we have only considered the so-called *weak interaction* between the sound and incident light, i.e., the interaction between the scattered light and sound has been ignored. In fact, scattered light may interact with the sound field again and produce higher orders of diffracted light. This rescattering process is characteristic of what is called *strong interaction.* (In the Bragg regime, for strong interaction, scattered light k_{+1} may rescatter back into the zeroth-order light). In the Raman-Nath regime, many orders may exist because plane waves of sound are available at the various angles required for scattering. The principle of the generation of many orders by rescattering is illustrated in Fig. 4.5. k_{+1} is generated through the diffraction of k_0 by K_{+1}, k_{+2} is generated through the diffraction of k_{+1} by K_{+2}, and so on, where the $K_{\pm p}$'s ($p = 0, \pm 1, \pm 2, \ldots$) denote the appropriate components of the plane wave spectrum of the sound. Again, the requirement of conservation of energy leads to the equation $\omega_m = \omega_0 \pm m\Omega$, with ω_m being the frequency of the m-th order scattered light.

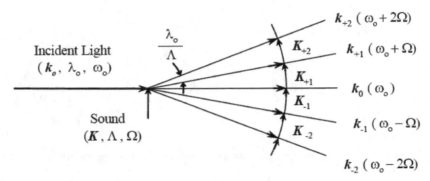

Fig. 4.5 Multiple scattering in the Raman-Nath regime.

4.2 The Acousto-Optic Effect: General Formalism

The interaction between the optical field $\mathcal{E}_0(\mathbf{R}, t)$ and sound field $S(\mathbf{R}, t)$ can be generally described by Maxwell's equations [Eqs. (2.1-1)-(2.1-4)]. We assume that the interaction takes place in an optically inhomogeneous, nonmagnetic isotropic medium, characterized by a permeability μ_0 and a permittivity $\tilde{\varepsilon}(\mathbf{R}, t)$ when a source-free optical field is incident on the time-varying permittivity. We write the time-varying permittivity as

$$\tilde{\varepsilon}(\mathbf{R}, t) = \varepsilon + \varepsilon'(\mathbf{R}, t), \tag{4.2-1}$$

where $\varepsilon'(\mathbf{R},t) = \varepsilon CS(\mathbf{R},t)$, i.e., it is proportional to the sound field amplitude $S(\mathbf{R},t)$, with C being the proportionality constant, dependent on the medium. Hence we can see that $\varepsilon'(\mathbf{R},t)$ represents the action of the sound field. The analysis presented below closely follows the work by Korpel (1972).

We will assume the incident light field $\boldsymbol{\mathcal{E}}_{inc}(\mathbf{R},t)$ satisfies Maxwell's equations with $\rho_v = 0$ and $\boldsymbol{J} = 0$. When the sound field interacts with $\boldsymbol{\mathcal{E}}_{inc}(\mathbf{R},t)$, the total field light fields $\boldsymbol{\mathcal{E}}(\mathbf{R},t)$ and $\boldsymbol{\mathcal{H}}(\mathbf{R},t)$ in the acoustic medium must also satisfy Maxwell's equations, rewritten as

$$\nabla \times \boldsymbol{\mathcal{E}}(\mathbf{R},t) = -\mu_0 \frac{\partial \boldsymbol{\mathcal{H}}(\mathbf{R},t)}{\partial t}, \tag{4.2-2}$$

$$\nabla \times \boldsymbol{\mathcal{H}}(\mathbf{R},t) = \frac{\partial}{\partial t} \left[\tilde{\varepsilon}(\mathbf{R},t) \boldsymbol{\mathcal{E}}(\mathbf{R},t) \right], \tag{4.2-3}$$

$$\nabla \cdot \left[\tilde{\varepsilon}(\mathbf{R},t) \boldsymbol{\mathcal{E}}(\mathbf{R},t) \right] = 0, \tag{4.2-4}$$

$$\nabla \cdot \boldsymbol{\mathcal{H}}(\mathbf{R},t) = 0. \tag{4.2-5}$$

Taking the curl of Eq. (4.2-2) and introducing it into Eq. (4.2-3), the equation for $\boldsymbol{\mathcal{E}}(\mathbf{R},t)$ becomes

$$\nabla \times \nabla \times \boldsymbol{\mathcal{E}}(\mathbf{R},t) = \nabla(\nabla \cdot \boldsymbol{\mathcal{E}}) - \nabla^2 \boldsymbol{\mathcal{E}}$$
$$= -\mu_0 \frac{\partial^2}{\partial t^2} [\tilde{\varepsilon}(\mathbf{R},t)\boldsymbol{\mathcal{E}}(\mathbf{R},t)]. \tag{4.2-6}$$

Now, from Eq. (4.2-4), we have

$$\nabla \cdot \tilde{\varepsilon}\boldsymbol{\mathcal{E}} = \tilde{\varepsilon}\nabla \cdot \boldsymbol{\mathcal{E}} + \boldsymbol{\mathcal{E}} \cdot \nabla \tilde{\varepsilon} = 0. \tag{4.2-7}$$

Assuming a two-dimensional $(x-z)$ sound field configuration with $\boldsymbol{\mathcal{E}}(\mathbf{R},t)$ polarized along the y-direction, i.e., $\boldsymbol{\mathcal{E}}(\mathbf{r},t) = E(\mathbf{r},t)\mathbf{a_y}$, we can readily show that $\boldsymbol{\mathcal{E}} \cdot \nabla \tilde{\varepsilon} = 0$. Hence Eq. (4.2-6) reduces to

$$\nabla^2 E(\mathbf{r},t) = \mu_0 \frac{\partial^2}{\partial t^2} [\tilde{\varepsilon}(\mathbf{r},t) E(\mathbf{r},t)], \qquad (4.2\text{-}8)$$

where r is the position vector in the $x - z$ plane. We will rewrite the term on the right-hand side of Eq. (4.2-8) as

$$\mu_0 \left[E \frac{\partial^2 \tilde{\varepsilon}}{\partial t^2} + 2 \frac{\partial E}{\partial t} \frac{\partial \tilde{\varepsilon}}{\partial t} + \tilde{\varepsilon} \frac{\partial^2 E}{\partial t^2} \right]. \qquad (4.2\text{-}9)$$

Because the time variation of $\tilde{\varepsilon}(\mathbf{r},t)$ is much slower than that of $E(\mathbf{r},t)$, i.e., the sound frequency is much lower than that of the light frequency, we will only retain the last term in Eq. (4.2-9) to get, using Eqs. (4.2-1) and (4.2-8),

$$\nabla^2 E(\mathbf{r},t) - \mu_0 \varepsilon \frac{\partial^2 E(\mathbf{r},t)}{\partial t^2} = \mu_0 \varepsilon' (\mathbf{r},t) \frac{\partial^2 E(\mathbf{r},t)}{\partial t^2}. \qquad (4\text{-}2\text{-}10)$$

Equation (4.2-10) is the scalar wave equation that is often used to investigate strong interaction in acousto-optics.

     We will now introduce harmonic variations in the incident light and sound in the forms

$$E_{inc}(\mathbf{r},t) = \mathrm{Re} \left[E_{inc}(\mathbf{r}) e^{j\omega_0 t} \right] = \frac{1}{2} E_{inc}(\mathbf{r}) e^{j\omega_0 t} + c.c., \qquad (4.2\text{-}11)$$

and

$$\frac{\varepsilon'(\mathbf{r},t)}{\varepsilon} = \mathrm{Re} \left[CS(\mathbf{r}) e^{j\Omega t} \right] = \frac{1}{2} CS(\mathbf{r}) e^{j\Omega t} + c.c., \qquad (4.2\text{-}12)$$

where c.c. denotes the complex conjugate. Note that $E_{inc}(\mathbf{r})$ and $S(\mathbf{r})$ in the above equations are phasor quantities. Because we have harmonic fields, we anticipate frequency mixing due to interaction. This is also evident from examining the right-hand side of Eq. (4.2-10). We will, therefore, cast the total field $E(\mathbf{r},t)$ into the form

$$E(\mathbf{r},t) = \frac{1}{2} \sum_{m=-\infty}^{\infty} E_m(\mathbf{r}) \exp\left[j(\omega_0 + m\Omega) t \right] + c.c. \qquad (4.2\text{-}13)$$

Substituting Eqs. (4.2-11)-(4.2-13) into Eq. (4.2-10) and assuming $\Omega \ll \omega_0$, we obtain, after some straightforward calculations, the following infinite coupled-differential equations:

$$\nabla^2 E_m(\mathbf{r}) + k_0^2 E_m(\mathbf{r}) + \frac{1}{2} k_0^2 CS(\mathbf{r}) E_{m-1}(\mathbf{r}) + \frac{1}{2} k_0^2 CS^*(\mathbf{r}) E_{m+1}(\mathbf{r}) = 0,$$
$$(4.2\text{-}14)$$

where $k_0 = \omega_0 \sqrt{\mu_0 \varepsilon}$ is the propagation constant of light in the medium and the asterisk denotes the complex conjugate. Note that $E_m(\mathbf{r})$ is the phasor amplitude of the m-th order light at frequency $\omega_0 + m\Omega$.

4.3 Raman-Nath Equations

We now consider a conventional interaction configuration shown in Fig. 4.6. A uniform sound wave of finite width L and propagating along x is often used, and we shall represent the sound phasor by

$$S(\mathbf{r}) = S(z,x) = Ae^{-jKx}, \qquad (4.3\text{-}1)$$

where A, in general, can be complex. Indeed, for $A = |A| e^{j\theta}$ and from Eq. (4.2-12), we see that $\varepsilon'(\mathbf{r},t) \sim C|A|\cos(\Omega t - Kx + \theta)$, which represents a plane wave of sound propagating along the x-direction, as shown in Fig. 4.6. We represent the incident plane wave of light as

$$E_{inc}(\mathbf{r}) = \psi_{inc} \exp\left(-jk_0 z \cos\phi_{inc} - jk_0 x \sin\phi_{inc}\right), \qquad (4.3\text{-}2)$$

where φ_{inc} is the incident angle as shown in Fig. 4.6. We look for a solution of $E_m(\mathbf{r})$ of the form

$$E_m(\mathbf{r}) = E_m(z,x) = \psi_m \exp\left(-jk_0 z \cos\phi_m - jk_0 x \sin\phi_m\right), \qquad (4.3\text{-}3)$$

with the choice for ϕ_m given by Eq. (4.1-1),

$$\sin \phi_m = \sin \phi_{inc} + m\lambda_0 / \Lambda = \sin \phi_{inc} + mK / k_0. \quad (4.3\text{-}4)$$

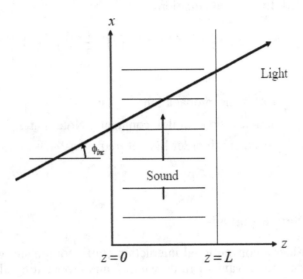

Fig. 4.6 Conventional sharp boundary interaction configuration.

Substituting Eqs. (4.3-1), (4.3-3), and (4.3-4) into Eq. (4.2-14), we obtain, after some considerable algebra,

$$\frac{\partial^2 \psi_m}{\partial x^2} - 2 j k_0 \sin \phi_m \frac{\partial \psi_m}{\partial x} - 2 j k_0 \cos \phi_m \frac{\partial \psi_m}{\partial z}$$

$$+ \frac{1}{2} k_0^2 CA^* \psi_{m+1} \exp\left[-j k_0 z \left(\cos \phi_{m+1} - \cos \phi_m \right)\right]$$

$$+ \frac{1}{2} k_0^2 CA \psi_{m-1} \exp\left[-j k_0 z \left(\cos \phi_{m-1} - \cos \phi_m \right)\right] = 0.$$

$$(4.3\text{-}5)$$

In deriving Eq. (4.3-5), we have assumed as usual that ψ_m is a slowly-varying function of z so that, within a wavelength of light, $\partial \psi_m / \partial z$ does not change appreciably; hence, $\partial^2 \psi_m / \partial z^2$ can be neglected when compared to $(2jk_0 \cos\phi_m) \partial \psi_m / \partial z$. In addition, according to physical intuition, we have used $\omega_0 \gg m\Omega$.

The physics behind Eq. (4.3-5) can be understood as follows: (a) the evolution of the m-th order scattered light (the third term on the left-hand side of the equation) in z is coupled to that of its adjacent orders ($m+1$ and $m-1$, see the last two terms), (b) the first term on the left-hand side of the equation is responsible for diffraction, and (c) the second term is merely the effect of the m-th order light traveling in a direction slightly different from z.

Often, the effect of propagational diffraction can be neglected if the width L of the cell is not too large. Also, because ϕ_m is a very small quantity (if $\phi_{inc} \ll 1$), we can assume $\psi_m(z,x) \approx \psi_m(z)$. Hence, Eq. (4.3-5) becomes

$$\frac{d\psi_m}{dz} = -j \frac{k_0 CA}{4\cos\phi_m} \psi_{m-1} \exp\left[-jk_0 z \left(\cos\phi_{m-1} - \cos\phi_m\right)\right]$$

$$-j \frac{k_0 CA^*}{4\cos\phi_m} \psi_{m+1} \exp\left[-jk_0 z \left(\cos\phi_{m+1} - \cos\phi_m\right)\right], \qquad (4.3\text{-}6)$$

which has to be solved with the boundary conditions

$$\psi_m = \psi_{inc} \delta_{m0} \quad \text{at } z \le 0, \qquad (4.3\text{-}7)$$

where δ_{m0} represents the Kronecker delta function. Equation (4.3-6) is one variant of the well-known *Raman-Nath equations* [Raman and Nath, 1935].

The physical interpretation of Eq. (4.3-6) is that there is a mutual coupling between neighboring orders in the interaction, that is, ψ_m is being contributed to by ψ_{m-1} and ψ_{m+1}. However, the phase of the contributions varies with z, and the exponents with arguments $k_0 z (\cos\phi_{m-1} - \cos\phi_m)$ and $k_0 z (\cos\phi_{m+1} - \cos\phi_m)$ represent the lack of phase synchronism in this coupling process. The lack of this phase synchronism will contribute little net interaction in the general case. However, there are two experimental configurations in which strong phase synchronisms occur. The first one being that the interaction length L is short enough such that the accumulated degree of phase mismatch is small. This is called the *Raman-Nath regime* or *Debye-Sears regime* and is characterized by the simultaneous generation of many diffracted orders as shown in Fig. 4.5. For the second configuration, called the *Bragg regime*, there exists phase synchronism between the 0-th order and the 1-st order or the 0-th order and the +1-st order. We shall now discuss the two configurations in the following sections.

4.4 Contemporary Approach

In this section, we re-formulate the Raman-Nath equations [Eq. (4.3-6)] into a set of equations extensively used by Korpel and Poon in their contemporary investigation of acousto-optics [Korpel and Poon (1980)].

Let us relate first the term CA to the refractive index variation $\Delta n(\mathbf{r},t)$ in the acousto-optic cell. Because

$$\tilde{\varepsilon}(\mathbf{r},t) = \varepsilon_0 n^2(\mathbf{r},t)$$

$$= \varepsilon_0 \left[n_0 + \Delta n(\mathbf{r},t) \right]^2$$

$$\cong \varepsilon_0 n_0^2 \left[1 + \frac{2\Delta n(\mathbf{r},t)}{n_0} \right], \qquad (4.4\text{-}1)$$

we compare Eq. (4.4-1) with Eq. (4.2-1) to obtain

$$CS(\mathbf{r},t) = \frac{2\Delta n(\mathbf{r},t)}{n_0}$$

or $C|S| = C|A| = 2(\Delta n)_{max} / n_0$, where $(\Delta n)_{max}$ denotes the peak amplitude of the assumed harmonic variation of $\Delta n(\mathbf{r},t)$. Assuming A is real and non-negative, i.e., $A = |A|$, the quantity $k_0 CAL / 2 = k_0 CA^* L / 2$ then can be written as

$$\frac{k_0 C |A| L}{2} = \left(\frac{k_0}{n_0}\right)(\Delta n)_{max} L = \alpha, \qquad (4.4\text{-}2)$$

where α represents the *peak phase delay* of the light through the acoustic medium.

Assuming small values of K/k_0, we expand the phase-asynchronism terms $k_0 z (\cos\phi_{m-1} - \cos\phi_m)$ and $k_0 z (\cos\phi_{m+1} - \cos\phi_m)$ in Eq. (4.3-6) in a power series by using Eq. (4.3-4),

$$k_0 z (\cos\phi_{m-1} - \cos\phi_m)$$

$$= k_0 z \left[\left(\frac{K}{k_0}\right)\sin\phi_{inc} + \left(m - \frac{1}{2}\right)\left(\frac{K}{k_0}\right)^2 + \dots\right], \qquad (4.4\text{-}3)$$

and

$$k_0 z (\cos\phi_{m+1} - \cos\phi_m)$$

$$= k_0 z \left[-\left(\frac{K}{k_0}\right)\sin\phi_{inc} - \left(m + \frac{1}{2}\right)\left(\frac{K}{k_0}\right)^2 + \dots\right]. \qquad (4.4\text{-}4)$$

We now define $\xi = z / L$, which is the normalized distance inside the acousto-optic cell, and $\xi = 1$ signifies the exit plane of the cell. We define one more important variable called the Q parameter or the Klien-Cook parameter in acousto-optics [Klein and Cook (1967)]:

$$Q = \frac{K^2 L}{k_0} = 2\pi L \frac{\lambda_0}{\Lambda^2}. \qquad (4.4\text{-}5)$$

Finally, with Eqs. (4.4-2), (4.4-3) and (4.4-4) by keeping their first two terms, and Eq. (4.4-5), we can re-write Eq. (4.3-6) as

$$\frac{d\psi_m}{d\xi} = -j\frac{\alpha}{2}e^{-j\frac{1}{2}Q\xi\left[\frac{\phi_{inc}}{\phi_B}+(2m-1)\right]}\psi_{m-1} - j\frac{\alpha}{2}e^{j\frac{1}{2}Q\xi\left[\frac{\phi_{inc}}{\phi_B}+(2m+1)\right]}\psi_{m+1},$$

(4.4-6)

where $\psi_m = \psi_{inc}\delta_{m0}$ at $\xi = 0$ from Eq. (4.3-7). In writing the above equation we have made small-angle approximations, i.e., $\sin\theta \approx \theta$, $\cos\theta \approx 1$, or $\phi_m \ll 1$. Equation (4.4-6) is a special case of the general plane wave multiple scattering theory, called the *Korpel-Poon multiple plane-wave scattering theory* [Korpel and Poon (1980), Appel and Somekh (1993)], which is valid for any sound field, not just a sound column as being considered here. We shall use these equations, called the *Korpel-Poon equations* [Poon and Kim (2005)] to investigate the two important conventional experimental configurations (the Raman-Nath regime and the Bragg regime) mentioned in Section 4.3.

4.5 Raman-Nath Regime

As mentioned in Eq. (4.1-6), the Raman-Nath regime is defined by the condition that $L \ll \Lambda^2 / \lambda_0$. Expressing this condition in terms of Q we have, using Eq. (4.4-5), $Q \ll 1$. Using this criterion, we see that the phase terms in the exponents in Eq. (4.4-6) involving $(m-1/2)Q$ and $(m+1/2)Q$ can be neglected if $mQ \ll 1$. Indeed, the condition $Q \ll 1$ is more stringent the more diffracted orders (m) are considered. Hence, for oblique incidence $(\phi_{inc} \neq 0)$, Eq. (4.4-6) becomes

$$\frac{d\psi_m}{d\xi} = -j\frac{\alpha}{2}e^{-j\frac{1}{2}Q\xi\frac{\phi_{inc}}{\phi_B}}\psi_{m-1} - j\frac{\alpha}{2}e^{j\frac{1}{2}Q\xi\frac{\phi_{inc}}{\phi_B}}\psi_{m+1}. \qquad (4.5-1)$$

For perpendicular incidence $(\phi_{inc} = 0)$, which is the first case treated historically, Eq. (4.5-1) becomes

$$\frac{d\psi_m}{d\xi} = -j\frac{\alpha}{2}(\psi_{m-1} + \psi_{m+1}). \tag{4.5-2}$$

Now recall the recursion relation for Bessel functions,

$$\frac{dJ_m(x)}{dx} = \frac{1}{2}\left[J_{m-1}(x) - J_{m+1}(x)\right]. \tag{4.5-3}$$

Then, writing $\psi_m = (-j)^m \psi'_m$, where $\psi'_m = J_m(\alpha\xi)$, we recognize, with $\psi_m = \psi_{inc}\delta_{m0}$ at $\xi = 0$, that the amplitude of the various scattered orders is

$$\psi_m = (-j)^m \psi_{inc} J_m(\alpha\xi). \tag{4.5-4}$$

Equation (4.5-4) is the well-known Raman-Nath solution. There are several important Bessel-function properties listed as follows:

1. $J_m(x)$ are real valued,

2. $J_m(x) = J_{-m}(x)$, for m even,

3. $J_m(x) = -J_{-m}(x)$, for m odd,

4. $\displaystyle\sum_{m=-\infty}^{\infty} J_m^2(x) = 1.$

Using property#4, we can show that at the exit of the sound column, i.e., $\xi = 1$, the sum of all the intensities of the diffracted light equals the incident intensity:

$$\sum_{m=-\infty}^{\infty} I_m = \sum_{m=-\infty}^{\infty} |\psi_m|^2 = |\psi_{inc}|^2 \sum_{m=-\infty}^{\infty} J_m^2(\alpha) = |\psi_{inc}|^2 = I_{inc}.$$

The above equation satisfies the conservation of energy.

4.6 Bragg Regime

Ideal Bragg diffraction is characterized by the generation of two scattered orders. For downshifted interaction $(\phi_{inc} = \phi_B)$, we have diffracted orders 0 and -1, whereas for upshifted interaction $(\phi_{inc} = -\phi_B)$, we have orders 0 and 1. According to Eq. (4.4-6), we have the following coupled equations describing the downshifted interaction:

$$\frac{d\psi_0}{d\xi} = -j\frac{\alpha}{2}\psi_{-1}, \tag{4.6-1}$$

and

$$\frac{d\psi_{-1}}{d\xi} = -j\frac{\alpha}{2}\psi_0. \tag{4.6-2}$$

Similarly, the coupled equations

$$\frac{d\psi_0}{d\xi} = -j\frac{\alpha}{2}\psi_1, \tag{4.6-3}$$

and

$$\frac{d\psi_1}{d\xi} = -j\frac{\alpha}{2}\psi_0 \tag{4.6-4}$$

describe the upshifted interaction.
Physically, we see that in downshifted interaction for example, according to Eq. (4.3-6), we have

$$\frac{d\psi_0}{dz} = \frac{-jk_0 CA}{4\cos\phi_0}\psi_{-1}\exp\left[-jk_0 z\left(\cos\phi_{-1} - \cos\phi_0\right)\right] \tag{4.6-5}$$

and

$$\frac{d\psi_{-1}}{dz} = \frac{-jk_0 CA^*}{4\cos\phi_{-1}} \psi_0 \exp\left[+jk_0 z\left(\cos\phi_{-1} - \cos\phi_0\right)\right] . \tag{4.6-6}$$

There exist phase synchronism between the 0-th and the -1-st orders when $\cos\phi_{-1} = \cos\phi_0$, as $\phi_{-1} = -\phi_0$. Referring to Fig. 4.7 (left), we have $\phi_0 = \phi_B = \phi_{inc}$, and $\phi_{-1} = -\phi_B = -\phi_{inc}$. Thus, the 0-th and the -1-st orders propagate symmetrically with respect to the sound wavefronts. Under these conditions, Eqs. (4.6-5) and (4.6-6) become Eqs. (4.6-1) and (4.6-2), respectively. Similar arguments may be advanced for the upshifted interaction case as shown in Fig. 4.7 (right).

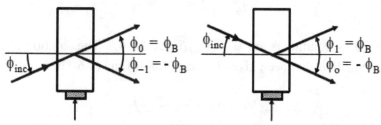

Fig. 4.7 Left: Downshifted Bragg diffraction ($\phi_{inc} = \phi_B$); Right: Upshifted Bragg diffraction ($\phi_{inc} = -\phi_B$). The convention for angles is counterclockwise positive.

The solutions to Eqs. (4.6-1) and (4.6-4), taking the boundary conditions Eq. (4.3-7) into account, read

$$\psi_0 = \psi_{inc} \cos\left(\alpha\xi / 2\right), \tag{4.6-7a}$$

$$\psi_{-1} = -j\psi_{inc} \sin\left(\alpha\xi / 2\right), \tag{4.6-7b}$$

for the downshifted interaction, and

$$\psi_0 = \psi_{inc} \cos\left(\alpha\xi / 2\right), \tag{4.6-8a}$$

$$\psi_1 = -j\psi_{inc} \sin\left(\alpha\xi / 2\right), \tag{4.6-8b}$$

for the upshifted case. Equations (4.6-7) and (4.6-8) are the well-known expressions for the scattered light in Bragg diffraction.

===

Conservation of Energy Example

The coupled equations for ideal Bragg diffraction are given by Eqs. (4.6-3) and (4.6-4) as follows:

$$\frac{d\psi_0}{d\xi} = -j\frac{\alpha}{2}\psi_1, \tag{a}$$

and

$$\frac{d\psi_1}{d\xi} = -j\frac{\alpha}{2}\psi_0. \tag{b}$$

The boundary conditions are $\psi_m = \psi_{inc}\delta_{m0}$ at $\xi \leq 0$. The solution of the coupled equations has been given in Eqs. (4.6-8a) and (4.6-8b) as follows:

$$\psi_0 = \psi_{inc}\cos(\alpha\xi/2),$$
$$\psi_1 = -j\psi_{inc}\sin(\alpha\xi/2).$$

It is clear that the conservation of energy is observed as

$$I_0 + I_1 = \psi_0\psi_0^* + \psi_1\psi_1^* = \psi_{inc}\psi_{inc}^* = I_{inc}.$$

However, the conservation of energy can be directly verified from the coupled equations without the need to find the solution. This is useful as it will give us some indication the derived coupled equations are appropriate. Let us calculate the following:

$$\frac{d}{d\xi}[I_0 + I_1] = \frac{d}{d\xi}[\psi_0\psi_0^* + \psi_1\psi_1^*]$$

$$= \frac{d\psi_0}{d\xi}\psi_0^* + \frac{d\psi_0^*}{d\xi}\psi_0 + \frac{d\psi_1}{d\xi}\psi_1^* + \frac{d\psi_1^*}{d\xi}\psi_1 .$$

Now substituting Eqs. (a) and (b) into the above, we have

$$\frac{d}{d\xi}[I_0 + I_1]$$

$$= -j\frac{\alpha}{2}\psi_1\psi_0^* + \left(-j\frac{\alpha}{2}\psi_1\right)^*\psi_0 + \left(-j\frac{\alpha}{2}\psi_0\right)\psi_1^* + \left(-j\frac{\alpha}{2}\psi_0\right)^*\psi_1$$

$$= 0,$$

which demonstrates the law of conservation of energy as the sum of the two diffracted intensities is constant along ξ.

===

When the acousto-optic modulator operates in the Bragg regime, the modulator is often called the *Bragg cell*. Figure 4.8 shows a typical Bragg cell operating at 40 MHz. In the figure, we denote the two diffracted laser spots at the far background. The incident laser beam (not visible as it traverses across a transparent medium of glass) is traveling along the long dimension of the transducer through the glass. The Bragg cell shown, model AOM-40, is commercially available [IntraAction Corporation]. It uses dense flint glass as an acoustic medium (refractive index $n_0 \sim 1.65$) and operates at sound center frequency of 40 MHz with a piezoelectric transducer height of about 2 mm (which we have neglected in our 2-D analysis) and an interaction length L of about 60 mm. Therefore, in the figure shown, the sound wave travels in the glass from the left to the right at velocity $V_S \sim 4000m/s$ with sound wavelength $\Lambda = V_S / f_S \sim 0.1\ mm$. If a He-Ne laser is used (its wavelength is about $0.6328\mu m$ in air), its wavelength inside glass is $\lambda_0 \sim 0.6328\mu m / n_0 \sim 0.3743\mu m$. Hence, the Bragg angle, according to

Eq. (4.1-2), inside the acoustic medium is $\sim 1.9 \times 10^{-3}$ radian or about 0.1 degree. For the parameters used and according to Eq. (4.4-5), we have a $Q \sim 14$.

For $\phi_{inc} = -(1+\delta)\,\phi_B$, where δ represents the deviation of the incident plane wave away from the Bragg angle, and limiting ourselves to orders 0 and 1 for upshifted interaction, Eq. (4.4-6) is reduced to the following set of coupled differential equations:

$$\frac{d\psi_0}{d\xi} = -j\frac{\alpha}{2}e^{-j\delta Q\xi/2}\psi_1, \qquad (4.6\text{-}9a)$$

and

$$\frac{d\psi_1}{d\xi} = -j\frac{\alpha}{2}e^{j\delta Q\xi/2}\psi_0 \qquad (4.6\text{-}9b)$$

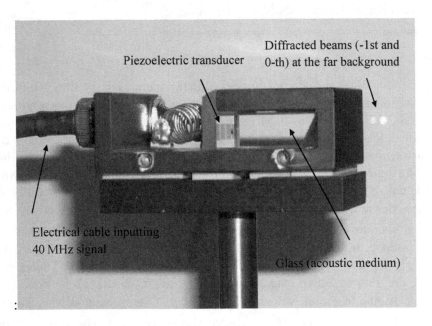

Fig. 4.8 Typical Bragg cell operating at 40 MHz [Adapted from Poon (2002)].

with the boundary conditions $\psi_0(\xi = 0) = \psi_{inc}$, and $\psi_1(\xi = 0) = 0$. Equation (4.6-9) may be solved analytically by a variety of techniques and the solutions are given by the well-known Phariseau formula [Phariseau (1956)] :

$$\psi_0(\xi) = \psi_{inc} e^{-j\delta Q\xi/4} \left\{ \cos\left[(\delta Q/4)^2 + (\alpha/2)^2\right]^{\frac{1}{2}} \xi \right.$$

$$\left. +j\frac{\delta Q}{4}\frac{\sin\left[(\delta Q/4)^2 + (\alpha/2)^2\right]^{\frac{1}{2}} \xi}{\left[(\delta Q/4)^2 + (\alpha/2)^2\right]^{\frac{1}{2}}} \right\}, \qquad (4.6\text{-}10a)$$

and

$$\psi_1(\xi) = \psi_{inc} e^{j\delta Q\xi/4} \left\{ -j\frac{\alpha}{4}\frac{\sin\left[(\delta Q/4)^2 + (\alpha/2)^2\right]^{\frac{1}{2}} \xi}{\left[(\delta Q/4)^2 + (\alpha/2)^2\right]^{\frac{1}{2}}} \right\}. \qquad (4.6\text{-}10b)$$

Equations (4.6-10a) and (4.6-10b) are similar to the standard two-wave solutions found by Aggarwal (1950) and adapted by Kogelnik (1969) to holography. More recently, it has been re-derived with the *Feynman diagram* technique [Chatterjee, Poon and Sitter, Jr. (1990)]. Note that by letting $\delta = 0$ we can reduce them to the solutions for ideal Bragg diffraction given by Eq. (4.6-8).

To show explicitly the effect of δ, we write down the weak interaction $(\alpha \to 0)$ version of Eq. (4.6-10) at $\xi = 1$. From Eq. (4.6-10b) and using Eq. (4.4-5), we have

$$\psi_1(\xi = 1) \propto \psi_{inc} e^{j\delta Q/4} \left\{ -j\frac{\alpha}{2}\frac{\sin(\delta Q/4)}{(\delta Q/4)} \right\}$$

$$= \psi_{inc} e^{j\delta Q/4} \left\{ -j\frac{\alpha}{2}\text{sinc}(\delta\phi_B L/\Lambda) \right\}. \qquad (4.6\text{-}11)$$

The sinc(.) term describes the angular plane wave spectrum of a transducer with width L. If we monitor the power in the first-order light and plot it as a function of the incident angle $\delta \times \phi_B$, we can trace the radiation pattern of the transducer. Figure 4.9(a) shows the experimental data points closely follow the theoretical radiation pattern of the AOM40 transducer, which is $\text{sinc}^2\left(\delta\phi_B L / \Lambda\right)$. Since the angles have been measured outside the acoustic medium, the first zero of sinc^2 (.) contains n_0 in the expression in Fig. 4.9(a). A physical interpretation of Fig. 4.9(a) is illustrated in Fig. 4.9(b). The incident plane wave of light is marked with an arrow at $-\phi_B$ as $\phi_{inc} = -\phi_B$. The incident light interacts with the sound plane wave from which it is offset by the Bragg angle. Therefore, the first-order diffraction is generated by the plane wave of sound at angle equal to 0, i.e., it is interacting with the plane wave of sound propagating normal to the transducer. Hence, the diffracted plane wave of light, marked by an arrow at ϕ_B, is proportional to the amplitude of this plane wave of sound. By increasing the angle of incidence, that is, moving the incident light arrow of Fig. 4.9(b) towards the right, the first-order light intensity will vary in proportion to the magnitude of the angular spectrum of sound [Cohen and Gordon (1965)]. It is therefore possible to explore acoustic radiation patterns by optical probing.

We have pointed out that the criterion for an acousto-optic modulator to operate in the Bragg regime is the condition that $Q \gg 1$. In physical reality, since a complete energy transfer between ψ_0 and $\psi_{\pm1}$ is never possible, there always exist more than two orders no matter how strong this condition becomes. This regime is generally known as the *near Bragg regime* [Poon and Korpel (1981a)] as $\left|\psi_0\right|^2 + \left|\psi_{\pm1}\right|^2 \neq \psi_{inc}^2$ due to the generation of higher orders. Now, the amount of first-order scattered light can be plotted as a function of Q at $\alpha = \pi$ and the Bragg regime is then defined arbitrarily by the condition that $\left|\psi_{\pm1}\right|^2 > 0.9\psi_{inc}^2$ (i.e., the *diffraction efficiency* for the first-order light is greater than 90%), which has been shown to translate to $Q > 7$ [Korpel (1972)]. For $Q \to \infty$, $\left|\psi_{\pm1}\right|^2 \to \psi_{inc}^2$ as expected. Acousto-optic diffraction in the near-Bragg (typically $Q \leq 2\pi$), and low-Q $\left(Q \to 0\right)$ regimes was examined numerically for the purpose of illustrating spurious orders, and the occasionally non-negligible amounts of scattered energies in them [Chen and Chatterjee (1996)].

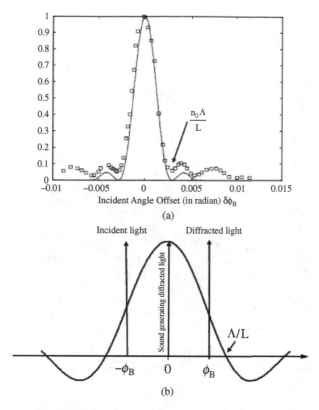

Fig. 4.9 (a) Angular plane wave spectrum of sound column: solid line theoretical with □ plotted as experimental data points [Adapted from Gies and Poon, 2002], (b) Physical interpretation of (a).

4.7 Numerical Examples

In this section, we shall use the Korpel-Poon equations, Eq. (4.4-6), to demonstrate some numerical results obtained by MATLAB®. In Table 4.1, we show the m-file for the calculation of diffracted intensities involving 10 diffracted orders in upshifted Bragg interaction. In using the m-file, we have created a MATLAB® function, AO_B10, which is shown in Table 4.2. The 10 coupled differential equations are shown as follows:

$$\frac{d\psi_5}{d\xi} = -j\frac{\alpha}{2}e^{-j\frac{1}{2}Q\xi\left[\frac{\phi_{inc}}{\phi_B}+9\right]}\psi_4 - 0$$

$$\frac{d\psi_4}{d\xi} = -j\frac{\alpha}{2}e^{-j\frac{1}{2}Q\xi\left[\frac{\phi_{inc}}{\phi_B}+7\right]}\psi_3 - j\frac{\alpha}{2}e^{j\frac{1}{2}Q\xi\left[\frac{\phi_{inc}}{\phi_B}+9\right]}\psi_5$$

$$\frac{d\psi_3}{d\xi} = -j\frac{\alpha}{2}e^{-j\frac{1}{2}Q\xi\left[\frac{\phi_{inc}}{\phi_B}+5\right]}\psi_3 - j\frac{\alpha}{2}e^{j\frac{1}{2}Q\xi\left[\frac{\phi_{inc}}{\phi_B}+7\right]}\psi_4$$

$$\frac{d\psi_2}{d\xi} = -j\frac{\alpha}{2}e^{-j\frac{1}{2}Q\xi\left[\frac{\phi_{inc}}{\phi_B}+3\right]}\psi_2 - j\frac{\alpha}{2}e^{j\frac{1}{2}Q\xi\left[\frac{\phi_{inc}}{\phi_B}+5\right]}\psi_3$$

$$\frac{d\psi_1}{d\xi} = -j\frac{\alpha}{2}e^{-j\frac{1}{2}Q\xi\left[\frac{\phi_{inc}}{\phi_B}+1\right]}\psi_1 - j\frac{\alpha}{2}e^{j\frac{1}{2}Q\xi\left[\frac{\phi_{inc}}{\phi_B}+3\right]}\psi_2$$

$$\frac{d\psi_0}{d\xi} = -j\frac{\alpha}{2}e^{-j\frac{1}{2}Q\xi\left[\frac{\phi_{inc}}{\phi_B}-1\right]}\psi_0 - j\frac{\alpha}{2}e^{j\frac{1}{2}Q\xi\left[\frac{\phi_{inc}}{\phi_B}+1\right]}\psi_1$$

$$\frac{d\psi_{-1}}{d\xi} = -j\frac{\alpha}{2}e^{-j\frac{1}{2}Q\xi\left[\frac{\phi_{inc}}{\phi_B}-3\right]}\psi_{-1} - j\frac{\alpha}{2}e^{j\frac{1}{2}Q\xi\left[\frac{\phi_{inc}}{\phi_B}-1\right]}\psi_0$$

$$\frac{d\psi_{-2}}{d\xi} = -j\frac{\alpha}{2}e^{-j\frac{1}{2}Q\xi\left[\frac{\phi_{inc}}{\phi_B}-5\right]}\psi_{-2} - j\frac{\alpha}{2}e^{j\frac{1}{2}Q\xi\left[\frac{\phi_{inc}}{\phi_B}-3\right]}\psi_{-1}$$

$$\frac{d\psi_{-3}}{d\xi} = -j\frac{\alpha}{2}e^{-j\frac{1}{2}Q\xi\left[\frac{\phi_{inc}}{\phi_B}-7\right]}\psi_{-4} - j\frac{\alpha}{2}e^{j\frac{1}{2}Q\xi\left[\frac{\phi_{inc}}{\phi_B}-5\right]}\psi_{-2}$$

$$\frac{d\psi_{-4}}{d\xi} = 0 - j\frac{\alpha}{2}e^{j\frac{1}{2}Q\xi\left[\frac{\phi_{inc}}{\phi_B}-7\right]}\psi_{-3}, \qquad (4.7\text{-}1)$$

where we have used $-4 \leq m \leq 5$ in Eq. (4.4-6).

In the first example, we run Bragg_regime_10.m for d = 0 corresponding to the case of exact Bragg incidence for upshifted interaction as we have defined $\phi_{inc}/\phi_B = -(1+\delta)$ in Eq. (4.7-1), and δ is "d" in the m-file. We then take $Q = 100$ to approximate ideal Bragg diffraction. The results are shown in Fig. 4.10. The vertical axis is $|\psi_m|^2 / |\psi_{inc}|^2$. Basically, we notice a complete exchange of power between the 0-th order and the first-order in Fig. 4.10(a) and higher orders are basically not present. In Fig. 4.10(b), a deviation from the exact Bragg angle incidence ($\delta = 0.02$, i.e., a deviation from the Bragg angle is $0.02\phi_B$) will result an incomplete energy exchange between the two orders.

As a second example, we take $\delta = 0$ (i.e., d = 0 in the m-file) and $Q = 5$, corresponding to near-Bragg diffraction The results are shown in Fig. 4.11, which corresponds well to the established results [Klien and Cook, 1967]. In this case, we notice that the periodic power exchange between the two main orders is no longer complete, indicating the presence of higher orders in this case, as shown in the figure.

Similarly, we use Eq. (4.4-6) to investigate Raman-Nath diffraction. We shall restrict ourselves to 13 diffracted orders $(-6 \le m \le 6)$ for normal incidence, i.e., d = -1. A set of 13 coupled equations therefore can be written out. Tables 4.1 and 4.2 can then be modified accordingly. Figure 4.12 shows the dependence of various scattered orders α when $Q = 0$, illustrating the Bessel function dependence given by Eq. (4.5-4). Figure 4.13 illustrates non-ideal Raman-Nath diffraction for normal incidence for $Q = 1.26$. Note that in the figure the true zeros, which are the characteristic of the Bessel functions for the ideal case, disappear for large value of α. The results presented in Fig. 4.13 correspond fairly well with established results [Mertens, Hereman, and Ottoy (1985)].

Table 4.1 Bragg_regime_10.m: m-file for solving 10 coupled equations in Bragg diffraction.

```
%Bragg_regime_10.m
%Bragg regime involving 10 diffracted orders
clear
d=input('delta = - incident angle/ Bragg angle -1 (enter 0 for exact Bragg angle incidence)
= ?')

Q=input('Q = (K^2*L)/ko (enter a large number, say 100, to get close to ideal
Bragg diffraction) = ')

n=0;
```

```
for

al=0:0.01*pi:8

n=n+1;
        AL(n)=al;
        [nz,y]=ode45('AO_B10', [0 1], [0 0 0 0 0 1 0 0 0 0], [], d, al, Q) ;
        [M1 N1]=size(y(:,1));
        [M2 N2]=size(y(:,2));
        [M3 N3]=size(y(:,3));
        [M4 N4]=size(y(:,4));
        [M5    N5]=size(y(:,5));
                        [M6
        N4]=size(y(:,6));
        [M7 N7]=size(y(:,7));
        [M8 N8]=size(y(:,8));
        [M9 N9]=size(y(:,9));
        [M10 N10]=size(y(:,10));
        psn2(n)=y(M8,8);
        psn1(n)=y(M7,7);
        ps0(n)=y(M6,6);
        ps1(n)=y(M5,5);
        ps2(n)=y(M4,4);
I(n)=y(M1,1).*conj(y(M1,1))+y(M2,2).*conj(y(M2,2))+y(M3,3)*conj(y(M3,3)) ...
        +y(M4,4)*conj(y(M4,4))+y(M5,5)*conj(y(M5,5)) ...                   b
        +y(M6,6)*conj(y(M6,6))+y(M7,7)*conj(y(M7,7)) ...
    +y(M8,8)*conj(y(M8,8))+y(M9,9)*conj(y(M9,9))+y(M10,10)*conj(y(M10,10));
end

   figure(1)
   plot(AL, ps0.*conj(ps0), '-', AL,ps1.*conj(ps1), ':', ...
       AL, psn1.*conj(psn1),'-.' AL,
       ps2.*conj(ps2),'--')
   title('Bragg regime')
   xlabel('alpha')
   axis([0 8 -0.1 1.1])
   legend('0 order', '1 order', '-1 order', '2 order')
grid on
```

Table 4.2   AO_B10.m: m-file for creating MATLAB® Function AO_B10 for use in Bragg_regime_10.m presented in Table 4.1.

```
%AO_B10.m
function dy= AO_B10(nz,y,options,d,a,Q)
```

```
 dy=zeros(10,1); %a column vector
% -4<= m <=5
% d=delta
% nz=normalized z

%m=5  -> y(1)
%m=4  -> y(2)
%m=3  -> y(3)
%m=2  -> y(4)
%m=1  -> y(5)
%m=0  -> y(6)
%m=-1 -> y(7)
%m=-2 -> y(8)
%m=-3 -> y(9)
%m=-4 -> y(10)

dy(1)=-j*a/2*y(2)*exp(-j*Q/2*nz*(-(1+d)+9))  +                0             ;

dy(2)=-j*a/2*y(3)*exp(-j*Q/2*nz*(-(1+d)+7))  +  -j*a/2*y(1)*exp(j*Q/2*nz*(-
(1+d)+9));

dy(3)=-j*a/2*y(4)*exp(-j*Q/2*nz*(-(1+d)+5))  +  -j*a/2*y(2)*exp(j*Q/2*nz*(-
(1+d)+7));

dy(4)=-j*a/2*y(5)*exp(-j*Q/2*nz*(-(1+d)+3))  +  -j*a/2*y(3)*exp(j*Q/2*nz*(-
(1+d)+5));

dy(5)=-j*a/2*y(6)*exp(-j*Q/2*nz*(-(1+d)+1))  +  -j*a/2*y(4)*exp(j*Q/2*nz*(-
(1+d)+3));

dy(6)=-j*a/2*y(7)*exp(-j*Q/2*nz*(-(1+d)-1))  +  -j*a/2*y(5)*exp(j*Q/2*nz*(-
(1+d)+1));

dy(7)=-j*a/2*y(8)*exp(-j*Q/2*nz*(-(1+d)-3))  +  -j*a/2*y(6)*exp(j*Q/2*nz*(-(1+d)-
1));

dy(8)=-j*a/2*y(9)*exp(-j*Q/2*nz*(-(1+d)-5))  +  -j*a/2*y(7)*exp(j*Q/2*nz*(-(1+d)-
3));

dy(9)=-j*a/2*y(10)*exp(-j*Q/2*nz*(-(1+d)-7))  +  -j*a/2*y(8)*exp(j*Q/2*nz*(-(1+d)-
5));

dy(10)=          0                + -j*a/2*y(9)*exp(j*Q/2*nz*(-(1+d)-
7));

return
```

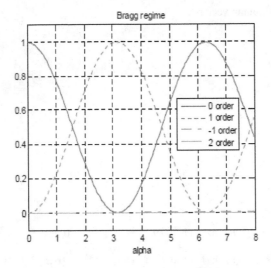

a)

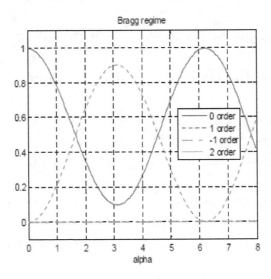

b)

Fig. 4.10 Intensity of diffracted orders versus peak phase delay α or "alpha" in the m-file for the ideal Bragg regime: a) Bragg angle incidence ($\delta = 0$), b) Near-Bragg angle incidence ($\delta = 0.02$).

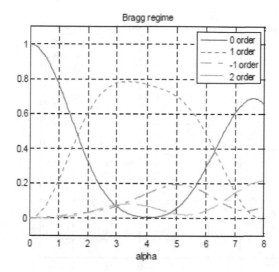

Fig. 4.11 Intensity of diffracted orders versus peak phase delay "alpha" for $Q = 5$.

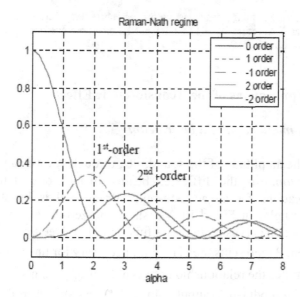

Fig. 4.12 Intensity of diffracted orders versus peak phase delay "alpha" in the ideal Raman-Nath regime $(Q = 0)$ (-1st and 1st orders are overlapped;-2nd and 2nd orders are overlapped).

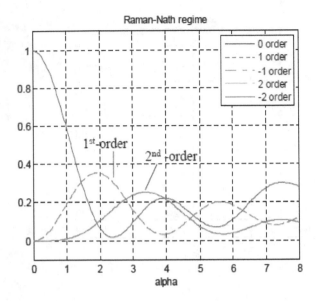

Fig. 4.13 Intensity of diffracted orders versus peak phase delay "alpha" in the non-ideal Raman-Nath regime ($Q = 1.26$)(-1st and 1st orders are overlapped;-2nd and 2nd orders are overlapped) .

4.8 Modern Applications of the Acousto-Optic Effect

4.8.1 Intensity modulation of a laser beam

By changing the amplitude of the sound, i.e., through α, we can achieve *intensity modulation* of the diffracted beams. In fact, one of the most popular applications for an acousto-optic Bragg cell is its ability to modulate laser light. In Fig. 4.14, we show the zeroth- and first-order intensity diffraction curves plotted as a function of α [see Eq. (4.6-8) for $\psi_{inc} = 1$], where P represents the bias point necessary for linear operation. Fig. 4.14 illustrates the relationship between the modulating signal, $m(t)$ and the intensity modulated output signal, $I(t)$. As shown in the figure, $m(t)$ is biased along the linear portion of the first order diffraction curve. It is clear that intensity modulation can also be achieved by using the zeroth diffracted order. However, the two diffracted orders process information in the linear regions with opposite slopes. This suggests that any demodulated electrical signal received via the first diffracted order will be 180 degrees out of phase with the electrical signal received via the zeroth diffracted order. Figure 4.15 shows an experimental setup used to demonstrate the principle. $m(t)$ is the modulating signal and the output of the amplitude-modulation (AM) modulator is an amplitude-modulated signal, $[b + m(t)]\cos(\Omega t)$, where b is adjusted to about $\alpha = \pi / 2$, and Ω is tuned to the center frequency of the acoustic transducer. When $m(t)$ is an triangular waveform, Fig. 4.16(a) shows the output of the two photodetectors PD0 and PD1 in Fig. 4.15. A phase shift of 180 degrees between the two electrical signals is clearly shown. In another experiment, $m(t)$ is an audio signal. Figure 4.16(b) shows the two 180 degrees out of phase detected electrical signals. With additional electronics and the introduction of feedback [to be discussed in Section 4.8.4], the acousto-optic system can be re-configured to produce pulse-width modulated optical signals [Poon, McNeill, and Moore (1997)].

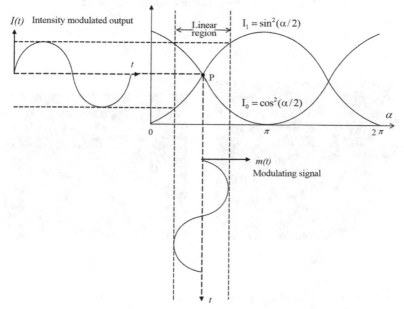

Fig. 4.14 Principle of acousto-optic intensity modulation system.

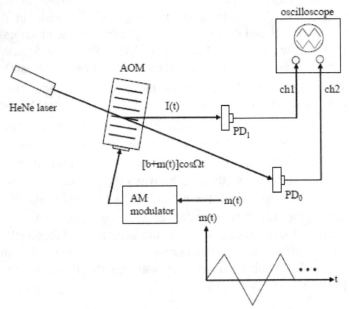

Fig. 4.15 Experimental setup for AM-demodulation.

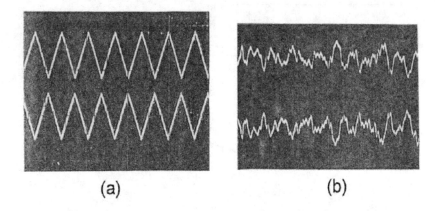

Fig. 4.16 Demodulated signals (a) triangular; (b) audio: Note that the upper traces (outputs from PD_1) are 180 degrees out of phase with the lower traces (outputs from PD_0) [After Poon *et al.* (1997)].

4.8.2 *Light beam deflector and spectrum analyzer*

In contrast to intensity modulation, where the amplitude of the modulating signal is varied, the frequency of the modulating signal is changed for applications in light deflection. Figure 4.17 shows a *light beam deflector* where the acousto-optic modulator operates in the Bragg regime. The angle between the first-order beam and the zeroth-order beam is defined as the *deflection angle* ϕ_d. We can express the change in the deflection angle $\Delta\phi_d$ upon a change $\Delta\Omega$ of the sound frequency as

$$\Delta\phi_d = \Delta(2\phi_B) = \frac{\lambda_0}{2\pi V_S}\Delta\Omega . \qquad (4.8\text{-}1)$$

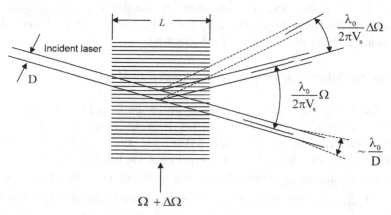

Fig. 4.17 Acousto-optic laser beam deflector.

Using a He-Ne laser $(\lambda_0 \sim 0.6\,\mu m)$, a change of sound frequency of 20 MHz around the center frequency of 40 MHz, and $V_S \sim 4\times10^3 m/s$, for the velocity of sound in glass, a change in deflection angle is $\Delta\phi_d \sim 3$ mrad. The *number of resolvable angles N* in such a device is determined by the ratio of the range of deflected angles $\Delta\phi_d$ to the angular spread of the scanning light beam. Since the angular spread of a beam of width D is of the order of λ_0 / D, we therefore have

$$N = \frac{\Delta\phi_d}{\lambda_0 / D} = \tau\frac{\Delta\Omega}{2\pi}, \qquad (4.8\text{-}2)$$

where $\tau = D / V_S$ is the *transit time* of the sound through the light beam. With the previously calculated $\Delta\phi_d \sim 3$ mrad and using a light beam of width $D = 5$ mm, the number of the achievable resolution is ~ 25 spots. Note that from Eq. (4.8-2) one can achieve improvement in resolution by expanding the lateral width of the light beam traversing the Bragg cell. Because the relation between the deflection angle and the frequency sweep is linear, a simple mechanism for high-speed laser beam scanning is possible through the acousto-optic effect because no moving mechanical parts are involved with this kind of scanning.

Instead of a single frequency input, the sound cell can be addressed simultaneously by a spectrum of frequencies. The Bragg cell scatters light beams into angles controlled by the spectrum of acoustic frequencies as each frequency generates a beam at a specific diffracted angle. Because the acoustic spectrum is identical to the frequency spectrum of the electrical signal being fed to the cell, the device essentially acts as a *spectrum analyzer* [see Problem 4.8].

4.8.3 *Demodulation of frequency-modulated (FM) signals*

From the preceding discussion, we recognize the Bragg cell's frequency-selecting capability. Here we discuss how to make use of this to demodulate frequency-modulated (FM) signals [Poon and Pieper (1983)], Pieper and Poon (1985)]. As seen from Fig. 4.18, the Bragg cell diffracts the light into angles ϕ_{di} controlled by the spectrum of carrier frequencies $\Omega_{0i}, i = 1, 2, \ldots,$ where each carrier has been frequency-modulated. For the i-th FM station, the instantaneous frequency of the signal is representable as $\Omega_i(t) = \Omega_{0i} + \Delta\Omega_i(t)$, where Ω_{0i} is a fixed carrier frequency and $\Delta\Omega_i(t)$ represents a time-varying frequency difference proportional to the amplitude of the modulating signal. As a usual practice, the FM variation $\Delta\Omega_i(t)$ is small compared to the carrier Ω_{0i}. Using Eq. (4.8-1) the i-th FM station is beamed, on the average, in a direction relative to the incident beam given by

$$\phi_{di} = \frac{\lambda_0 \Omega_{0i}}{2\pi V_S}.$$

$$(4.8\text{-}3)$$

This is illustrated in Fig. 4.18. For each FM carrier, there will now be an independently scattered light beam in a direction determined by the carrier frequency. For clarity, only a few of the scattered light beams are shown. The principle behind the *FM demodulation* is that the actual instantaneous angle of deflection deviates slightly from Eq. (4.8-3) due to the inclusion of $\Delta\Omega_i(t)$, which causes a "wobble" $\Delta\phi_{di}(t)$ in the deflected beam. In particular, we find, with Eq. (4.8-2),

$$\Delta\phi_{di}(t) = \left(\frac{\lambda_0}{2\pi V_s}\right)\Delta\Omega_i(t). \qquad (4.8\text{-}4)$$

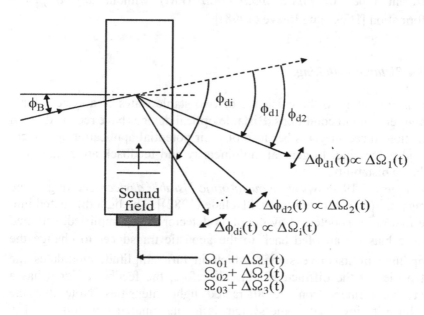

Fig. 4.18 Principle of acousto-optic FM demodulation.

Because, in FM, the frequency variation $\Delta\Omega_i(t)$ is proportional to the amplitude of the audio signal, the variation in the deflected angle $\Delta\phi_{di}(t)$ is likewise proportional to the modulating signal. By placing a knife-edge screen in front of a photodiode positioned along the direction of ϕ_{di}, the amount of light intensity reaching the photodiode, to first order varies linearly with the small wobble $\Delta\Omega_i(t)$ and hence gives a current proportional to $\Delta\Omega_i(t)$ [see Eq. (4.8-4)], that is, proportional to the amplitude of the modulating signal. In fact, by placing an array of knife-edge-screened detectors positioned along the various deflected beams, we can monitor all the FM stations simultaneously. This knife-edge technique has been used previously for surface acoustic wave detection [Whitman and Korpel (1969)]. The factors that limit the performance of acousto-optic FM demodulators have been investigated in the context of the use of a knife-edge detector as well as a bicell detector (two photo-active areas separated by a small gap) [Brooks and Reeve (1995)]. The use of a bicell detector has also been able to identify and demodulate a wide range of different types of *phase modulation* (PM) without any *a priori* information [Hicks and Reeve (1998)].

4.8.4 *Bistable switching*

Bistability refers to the existence of two stable states of a system for a given set of input conditions. Bistable optical devices have received much attention in recent years because of their potential application in optical signal processing. In general, nonlinearity and feedback are required to achieve bistability.

Figure 4.19 shows an *acousto-optic bistable device* operating in the Bragg regime [Chrostowski and Delisle (1982)]. The light diffracted into the first order is detected by the photodetector (PD), amplified, summed with a bias α_0, and fed back to the acoustic transducer to change the amplitude of its drive signal, which in turn amplitude-modulates the intensities of the diffracted light. Therefore, the feedback signal has a recursive influence on the diffracted light intensities. Note that the nonlinearity involved in the system is a sine-squared function [see Eq. (4.6-8b)]:

$$I_1 = |\psi_1|^2 = I_{inc}\sin^2(\alpha/2) \; , \tag{4.8-5}$$

where $I_{inc} = |\psi_{inc}|^2$ is the incident intensity and we have a system with a nonlinear input (α) -output (I_1) relationship.

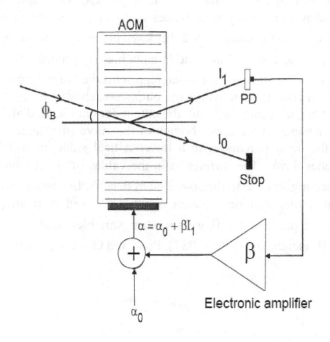

Fig. 4.19 Acousto-optic bistable device.

The effective α scattering the light in the acousto-optic cell is given by the *feedback equation*,

$$\alpha = \alpha_0 + \beta I_1, \qquad (4.8\text{-}6)$$

where β denotes the product of the gain of the amplifier and the quantum efficiency of the photodetector.

Note that under the feedback action, α can no longer, in general, be treated as a constant. In fact, α can be treated as constant during interaction if the interaction time, given as the ratio of the laser beam width and the speed of sound in the cell, is very small compared to the delays incorporated by the finite response time of the photodetector, the sound-cell driver and the feedback amplifier, or any other delay line that may be purposely installed (e.g., an optical fiber or coaxial cable) in the feedback path. We consider this case only. The steady-state behavior of the system is given by the simultaneous solution of Eqs. (4.8-5) and (4.8-6). Figure 4.20 shows the steady state values of I_1 for increments in α_0 for $\beta = 2.6$ and $Q \to \infty$ (meaning only 2 diffracted orders are involved). Note that, from Fig. 4.20, a gradual increase in input bias α_0 produces a steady increase in the output intensity I_1 (which represents the *lower stable state*) until reaching a critical value where the output switches up to the *higher stable state*. On decreasing the input, the output does not immediately fall sharply but remains on the upper branch of the curve (the higher stable state) until the input is reduced to a lower critical value, at which the output switches down. The difference in the values of α_0 at which the transitions occur gives rise to the so-called *hysteresis*. In passing, we also mention that similar hysteric behavior may be observed by treating I_{inc} or β as the input and treating the other variables such as α_0 as parameters. [Banerjee and Poon (1987), Poon and Cheung (1989)].

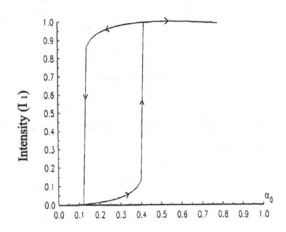

Fig. 4.20 Hysteresis by acousto-optics [Adapted from Poon *et al.*, (1997)].

==

Acousto-Optic Bistability Simulation Problem

We solve

$$I_1 = |\psi_1|^2 = I_{inc} \sin^2(\alpha / 2),$$

and

$$\alpha = \alpha_0 + \beta I_1,$$

simultaneously and plot the hysteresis loop for $\beta = 2.5$ using MATLAB. The simulation and its output are shown below.

```
clear; clc
I1=linspace(0,1,1013);
I1_out=linspace(0,1,1013);

I2=linspace(0,1,1013);
I2_out=linspace(0,1,1013);

I_inc=1;
alpha0=linspace(0,1,1013);
beta=2.5;

for n=1:1012
    I1_out(n)=I_inc*sin((alpha0(n)+beta*I1(n))/2).^2;
    I1(n+1)=I1_out(n);
end

I2(1012)=I1(1012);

for n=1012:-1:2
    I2_out(n)=I_inc*sin((alpha0(n)+beta*I2(n))/2).^2;
    I2(n-1)=I2_out(n);
end

End

figure(1)
plot(alpha0, I1_out)
```

```
hold on
plot(alpha0, I2_out)
xlabel('alpha_0')
ylabel('Intensity(I_1)')
```

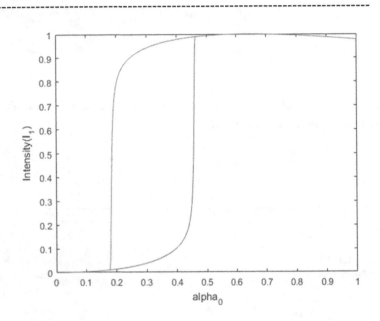

==

Figure 4.21(a) shows the experimental setup for the bistable system, where the scaling adder is an electronic summer. It sums the input from photodiode (PD_1) and the signal from the function generator to give the output to drive the radio-frequency (RF)-driver. The function generator is used to provide signals for the bias α_0. Figure 4.21(b) shows hysteresis involving the first order and the zeroth-order when AOM-40 was used. The investigation of optical bistability in the *second-order Bragg regime*, i.e., the incident angle is at twice the Bragg angle [Alferness (1976), Poon and Korpel (1981b)], has led to better performance in terms of wider hysteresis [McNeill and Poon (1992)]. Most recently, the use of an acousto-optic modulator with feedback has led to the investigation of an optical set-reset flip-flop [Chen and Chatterjee (1997)].

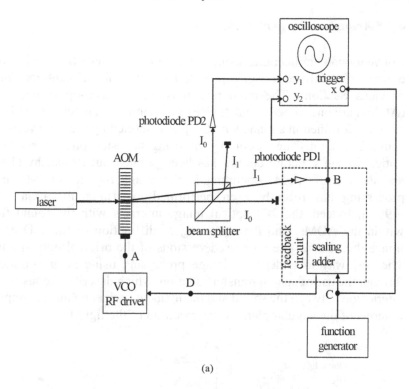

(a)

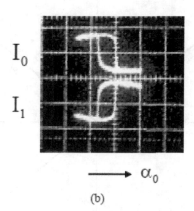

(b)

Fig. 4.21(a) Practical implementation of acousto-optic bistable system and (b) experimental results [Adapted from Poon and Cheung (1989)].

4.8.5 *Acousto-optic spatial filtering*

Conventional applications using acousto-optic interactions have been extensively confined to signal processing such as those mentioned in the previous Sections. The main reason is that acousto-optic modulators (AOMs) are one-dimensional devices as the interaction between light and sound is confined in a plane with the plane defined by the wavevectors of sound and light. The possibility of using acousto-optic interaction to optical image structure control has been pointed out [Balakshy (1984)] and the use of AOMs operating in the Bragg regime for 2-D image processing has recently been performed experimentally [Xia *et al.* (1996)]. Indeed, the 2-D optical image interacts with the sound fields within the AOMs, and the scattering or diffraction of the 2-D optical image then carries the processed versions of the original optical image. The best way to understand image processing using acousto-optics is through the use of spatial transfer function, which describes acousto-optic interaction between the sound and the incident optical image decomposed in terms of the angular plane wave spectrum of the light field.

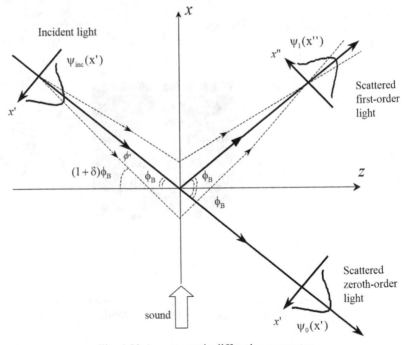

Fig. 4.22 Acousto-optic diffraction geometry.

Figure 4.22 shows how an arbitrary incident field propagates through a Bragg cell operating in the upshifted mode. Limiting ourselves to two diffracted orders and the case of zeroth-order light, we decompose the incident field into plane waves with different amplitudes propagating in direction defined by $\phi' = \delta \times \phi_B$. Equation (4.6-10) gives the plane wave amplitude of the zeroth-order light when a plane wave ψ_{inc} is incident at ϕ' away from the nominally Bragg angle of incidence. Hence we can define the so-called transfer function of the zeroth-order light as follows:

$$H_0(\delta) = \frac{\psi_0(\xi)\big|_{\xi=1}}{\psi_{inc}}, \qquad (4.8\text{-}7)$$

where $\psi_0(\xi)\big|_{\xi=1}$ is the result from Eq. (4.6-10) evaluated at the exit of the Bragg cell. This definition of the transfer function permits us to relate the input (incident) spectrum, $\Psi_{inc}(k_{x'})$, to the output (zeroth-order) spectrum, $\Psi_0(k_{x'})$, as

$$\Psi_0(k_{x'}) = \Psi_{inc}(k_{x'}) H_0(\delta), \qquad (4.8\text{-}8)$$

where $\psi_{inc}(x') = \mathcal{F}^{-1}\{\Psi_{inc}(k_{x'})\}$ and $\psi_0(x') = \mathcal{F}^{-1}\{\Psi_0(k_{x'})\}$ are the field distribution of the incident light and that of the zeroth-order light, respectively, $\mathcal{F}^{-1}$ denotes an inverse Fourier transform operation with x' and $k_{x'}$ representing the transform variables. Finally, the concept of spatial filtering becomes clear when we relate the spatial frequency $k_{x'}$ along x' to ϕ' [see Fig. 4.22] by

$$k_{x'} = k_0 \sin(\phi') \cong k_0 \phi' = k_0 \delta \phi_B = \pi \delta / \Lambda$$

as $\phi' = \delta \times \phi_B$ and $\phi_B = \lambda_0 / 2\Lambda$. Equation (4.8-8) can now be written in terms of spatial frequency if we use the above derived relationship such that $\delta = k_{x'} \Lambda / \pi$:

$$\Psi_0(k_{x'}) = \Psi_{inc}(k_{x'}) H_0(k_{x'} \Lambda / \pi) . \qquad (4.8\text{-}9)$$

The spatial distribution $\psi_0(x')$ is then given by

$$\psi_0(x') = \mathcal{F}^{-1}\{\Psi_{inc}(k_{x'})H_0(k_{x'}\Lambda/\pi)\},$$

$$= \frac{1}{2\pi}\int_{-\infty}^{\infty}\Psi_{inc}(k_{x'})H_0(k_{x'}\Lambda/\pi)\exp(-jk_{x'}x')dk_{x'}. \qquad (4.8\text{-}10)$$

Equation (4.8-10) determines the profile of the scattered zeroth-order field, $\psi_0(x')$, from any arbitrary incident field, $\psi_{inc}(x')$, in the presence of the acoustic field. The formalism in Eq. (4.8-10) is similar to the one developed by Magdich, Molchanov and Ya [1977], which was later developed by Chatterjee, Poon and Sitter, Jr. [1990] in the investigation of acousto-optic beam distortion.

   For $Q = 14$, $\alpha = \pi$ and $\Lambda = 0.1$ mm, Fig. 4.23(a) shows the magnitude plot of the zeroth-order transfer function $|H_0(k_x\Lambda/\pi)|$ [see Eq. (4.6-10a) with δ replaced by $k_x\Lambda/\pi$] as a function of k_x. For brevity, we have dropped the prime from the variable x' with the understanding that it is related to the x' coordinate. This figure makes physical sense. When a plane wave is incident at the exact Bragg angle, the plane wave is completely diffracted to the first order light as it is evident from the plot that when $k_x = 0$, the magnitude of the transfer function is zero. Now for incident angles that are far away from the exact Bragg angle incidence, that is, k_x is large, we expect that the incident plane wave will go through the Bragg cell without any interaction with the sound, which is evident that $|H_0| = 1$ at large values of k_x. The transfer function exhibits *highpass spatial filtering*. Figures 4.23(a) and (b) show the incident and the zeroth-order beams, respectively. Note that when the zeroth-order beam has been spatially filtered by the transfer function shown by Fig. 4.23(a), its edges have been emphasized. Table 4.3 shows the m-file for generating these simulations.

   We can explain the phenomenon of highpass filtering or edge enhancement in the zeroth-order diffracted light theoretically from the transfer function directly. Using the fact that $\delta = k_x\Lambda/\pi$ and from Eq. (4.6-10a) together with Eq. (4.8-7), we can express H_0 as

$$H_0\left(k_x \Lambda / \pi\right) \approx \exp\left(-jQ\Lambda k_x / 4\pi\right)\left[A + jBk_x\right] \approx \left[A + jBk_x\right] \quad (4.8\text{-}11)$$

if we assume that $\left(\delta_{\max} Q / 4\right) \ll \left(\alpha / 2\right)$, where $\delta_{\max}$ is the maximum extension of δ, $A = \cos\left(\alpha / 2\right)$ and $B = \dfrac{Q\Lambda}{4\pi} \dfrac{\sin\left(\alpha / 2\right)}{\left(\alpha / 2\right)}$. In addition, since the exponential term is related to the spatial position shift of the diffracted beam along the x-axis, it is, therefore, not important to the final image presentation, and hence not included in the final step of the approximation in Eq. (4.8-11). We can now interpret Eq. (4.8-11) in the spatial domain. From Table 2.2 of Chapter 2, we have the Fourier transform property

$$\mathcal{F}\left\{\partial f\left(x, y\right) / \partial x\right\} = -jk_x \mathcal{F}\left\{f\left(x, y\right)\right\},$$

where $\mathcal{F}$ again denotes a Fourier transform operation, and using Eqs. (4.8-9) and (4.8-10), the amplitude of the zeroth-order diffracted light written in a full 2-D version, $\psi_0\left(x, y\right)$, at the exit of the Bragg cell, can be approximately written as

$$\psi_0\left(x, y\right) = \left(A - B\frac{\partial}{\partial x}\right)\psi_{inc}\left(x, y\right).$$

For the incident field given by Figs. 4.23(b) and 4.23(c) illustrates the intensity of the diffracted light given by

$$\left|\psi_0\left(x, y\right)\right|^2 = \left|\left(A - B\frac{\partial}{\partial x}\right)\psi_{inc}\left(x, y\right)\right|^2.$$

Note that in Fig. (4.23c), it shows a pure differentiation result as $A = 0$ for $\alpha = \pi$. We see that under the condition $\left(\delta_{\max} Q / 4\right) \ll \left(\alpha / 2\right)$ and $\alpha = \pi$, the zeroth-order diffracted light performs first-order spatial derivative on the incident light profile and, therefore, mathematical operations such as differentiation can be achieved optically in real time

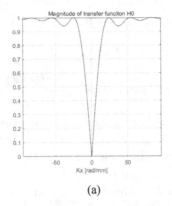

(a)

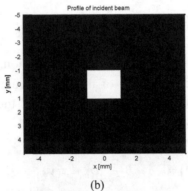

(b)

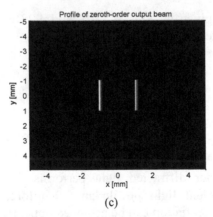

(c)

Fig. 4.23 a) Magnitude of zeroth-order transfer function $|H_0|$ as a function of spatial frequency $k_{x'}$ for $Q = 14$, $\alpha = \pi$ and $\Lambda = 0.1$ mm, b) Incident beam profile, c) Output zeroth-order beam profile.

In the above example, both edges of the square image have been extracted equally along the x-direction. Selective edge extraction, that is, either only the left edge of the image or the right edge of the image, can be performed when we change the sound pressure through α. Using the same m-file shown in Table 4.3, we change the peak phase delay, i.e., α, to $\alpha = 0.6\pi$. Figure 4.24(a) shows its transfer function and Fig. 4.24(b) shows that the left edge of the square has been emphasized. Indeed recent experiments have confirmed the idea experimentally [Davis and Nowak (2002)]. To obtain higher derivatives, we can, for example, have two acousto-optic modulators cascaded [Cao, Banerjee and Poon (1998)]. With two AOMs oriented orthogonally to each other, a second-order mixed derivative operation has also been demonstrated [Banerjee, Cao, and Poon (1997)]. Most recently, anisotropic acousto-optic diffraction has been explored for two-dimensional optical image processing [Voloshinov, Babkina, and Molchanov (2002); Balakshy, Voloshinov, Babkina and Kostyuk (2005)]. In passing, we point out that edge enhancement using volume phase gratings also has been reported [Case (1979), Marquez *et al*. (2003)].

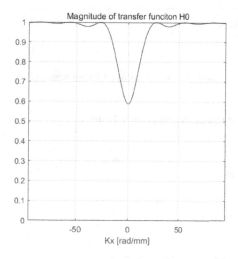

(a)

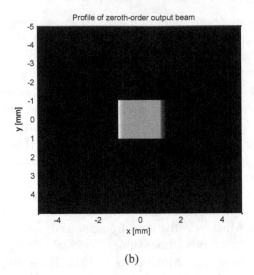

(b)

Fig. 4.24 a) Magnitude of zeroth-order transfer function $|H_0|$ as a function of spatial frequency $k_{x'}$ for $Q = 14$, $\alpha = 0.6\pi$ and $\Lambda = 0.1$ mm. b) Output zeroth-order beam profile, where input beam profile is shown in Fig. 4.23(b).

Table 4.3 AO_spatial_filtering.m.

```
%AO_spatial_filtering.m
clear
Q=input('Q =
'); z=1;
Ld=input('Wavelength of sound wave in [mm] (0.1 mm nominal value)= ');
Ld=Ld*10^-3;
al=input('alpha = ');
X_Sx=input('Length of square in [mm] (enter 1-3)= ');
X_Sx=X_Sx*10^-3;
Xmin=-0.005;
Xmax=0.005;
Step_s=0.5*6.5308e-005;
x=Xmin:Step_s:Xmax;

%Length of consideration range
L=Xmax-Xmin;

%Ratio of the square about the consideration range
R=X_Sx/L;

n=size(x);
N=n(2);
```

```
%Size of the square
Sx=round(N*R);

%Square
Einc=zeros(N,N);
S=size( Einc( round(N/2)-round(Sx/2):round(N/2)+round(Sx/2), ...
   round(N/2)-round(Sx/2):round(N/2)+round(Sx/2)));
Einc(round(N/2)-round(Sx/2):round(N/2)+round(Sx/2), ...
   round(N/2)-round(Sx/2):round(N/2)+round(Sx/2) )=ones(S);

%Square profile in Frequency domain using FFT
F_Einc=fft2(Einc);
F_Einc=fftshift(F_Einc);

%dx is the step size
dx=L/(N-1);

%Calcuation of range vector for space and frequency for
k=1:N

   X(k)=L/N*(k-1)-L/2;
   Y(k)=L/N*(k-1)-L/2;

   %Kx=(2*pi*k)/(N*dx)
   %k is sampling number, N is number of sample, dx is step size
   Kxp(k)=(2*pi*(k-1))/((N-1)*dx);
   Kyp(k)=(2*pi*(k-1))/((N-1)*dx);

end Kx=Kxp-
max(Kxp)/2; Ky=Kyp-
max(Kyp)/2;

%Calculation of  H0
for l=1:N
         for m=1:N

   d_x=Kx(l)*Ld/pi;
   D_x(l)=d_x;
   A_x=  cos( ( ( (d_x*Q/4)^2 + (al/2)^2 )^0.5 ) *z);
   Bdn_x= sin( ( ( (d_x*Q/4)^2 + (al/2)^2 )^0.5 ) *z);
   Bn_x=      ( (d_x*Q/4)^2 + (al/2)^2 )^0.5;
   B_x=Bdn_x/Bn_x;

   H0(m,l)=exp(-
j*d_x*Q*z/4)*(A_x+j*d_x*Q/4*B_x);
   end

end
```

```
%Calculation of profile of zero order output beam in Frequency domain
F_Eout_zero=F_Einc.*H0;
F_Eout_zero=fftshift(F_Eout_zero);

%Zeroth-order output beam profile in Space domain
E_out_zero=ifft2(F_Eout_zero);

%axis in [mm] scale
X=X*10^3;
Y=Y*10^3;
Kx=Kx*10^-3;
Ky=Ky*10^-3;

figure(1) plot(Kx, abs(H0(154,: )))
title('Magnitude of transfer funciton
H0') xlabel(' Kx [rad/mm] ')
grid on
axis([ min(Kx) max(Kx) 0 1])
axis square
figure(2)
image(X,Y,256*Einc/max(max(Einc)))
colormap(gray(256))
title('Profile of incident beam')
xlabel('x [mm]')
ylabel('y [mm]')
axis square
figure(3)
image(X,Y, 256*abs(E_out_zero)/max(max(abs(E_out_zero))))
colormap(gray(256))
title('Profile of zeroth-order output beam'
) xlabel('x [mm]')
ylabel('y [mm]')
axis square
```

4.8.6 *Acousto-optic heterodyning*

Optical information is usually carried by coherent light such as laser. With reference to Eq. (2.3-2), we let

$$\psi(x,y,z=0,t) = \psi_p(x,y;0)\exp(j\omega_0 t)$$

represent the coherent light field on the surface of the photodetector (PD). The light field is oscillating at temporal frequency ω_0. As the photodetector responds to intensity, i.e., $|\psi|^2$, it gives the current, i, as output by spatially integrating the intensity:

$$i \propto \int_D |\psi_p \exp(j\omega_0 t)|^2 \, dxdy, \tag{4.8-12}$$

$$= \int_D |\psi_p|^2 \, dxdy = A^2 D,$$

where D is the active surface area (area that is sensitive to light intensity) of the photodetector and the last step of the expression is obtained if we consider $\psi_p = A$, i.e., a uniform plane wave incidence. The situation is shown in Fig. 4.25.

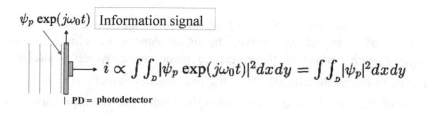

Fig. 4.25 Direct detection.

In this case, we can see that the photodetector current is proportional to the intensity, A^2, of the incident light. Therefore, the output current varies according to the intensity of the information signal. This mode of photodetection is called *direct detection* or *incoherent detection* in optics.

Let us now consider the *heterodyning* of two plane waves on the surface of the photodetector. The situation is shown in Fig. 4.26. We assume an information-carrying plane wave, also called the information signal plane wave, $A_s \exp\{j[\omega_0 t + s(t)]\} \exp(-jk_0 x \sin\phi)$, and a reference plane wave, $A_r \exp\left[j(\omega_0 + \Omega)t \right]$, or called a *local oscillator* as known in radio.

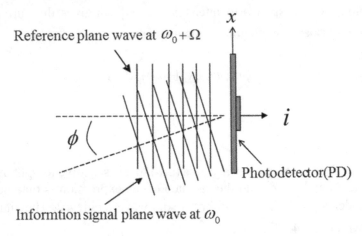

Reference plane wave at $\omega_0 + \Omega$

ϕ

i

Photodetector(PD)

Informtion signal plane wave at ω_0

Fig. 4.26 Heterodyning or coherent detection.

Note that the frequency of the reference plane wave is Ω higher than that of the signal plane wave. The signal plane wave is inclined at an angle ϕ with respect to the reference plane wave, which is incident normal to the photodetector. Also, the information content $s(t)$ is in the phase of the signal wave in this example. Under the situation shown in Fig. 4.26, we see that the two plane waves interfere on the surface of the photodetector, giving the total light field ψ_t as

$$\psi_t = A_r \exp\left[j(\omega_0 + \Omega)t \right] + A_s \exp\left\{ j\left[\omega_0 t + s(t) \right] \right\} \exp\left(-jk_0 x \sin\phi \right).$$

Again, the photodetector responds to intensity, giving the current

$$i \propto \int_D |\psi_t|^2 \, dxdy$$

$$= \int_{-a}^{a} \int_{-a}^{a} \left[A_r^2 + A_s^2 + 2A_r A_s \cos\left(\Omega t - s(t) + k_0 x \sin\phi \right) \right] dxdy,$$

where we have assumed that the photodetector has a uniform active area of $2a \times 2a$. The current can be evaluated to give

$$i(t) \propto 2a\left(A_r^2 + A_s^2\right) + 4A_r A_s \frac{\sin\left(k_0 a \sin\phi\right)}{k_0 \sin\phi}\cos\left(\Omega t - s(t)\right), \quad (4.8\text{-}13)$$

where we have assumed A_r and A_s are real for simplicity. The current output has two parts: the DC current and the AC current. The AC current at frequency Ω is known as the heterodyne current. Note that the information content $s(t)$ originally embedded in the phase of the signal plane wave has now been preserved and transferred to the phase of the heterodyne current. Indeed, this phase-preserving technique in optics is commonly known as *holographic recording* or *wavefront recording* [Gabor (1948)]. The above process is called *optical heterodyning* [Poon (2004)]. In optical communications, it is often referred to as *optical coherent detection*. In contrast, if the reference plane wave were not used for the detection, we have incoherent detection as previously described. The information content carried by the signal plane wave would be lost as it is evident that for $A_r = 0$, the above equation gives only a DC current at a value proportional to the intensity of the signal plane wave, A_s^2.

Let us now consider some of the practical issues encountered in heterodyning. Again, the AC part of the current given by the above equation is the heterodyne current $i_{het}(t)$ given by

$$i_{het}(t) \propto A_r A_s \frac{\sin\left(k_0 a \sin\phi\right)}{k_0 \sin\phi}\cos\left(\Omega t - s(t)\right). \quad (4.8\text{-}14)$$

We see that since the two plane waves propagate in slightly different directions, the heterodyne current output is degraded by a factor of $\sin\left(k_0 a \sin\phi\right)/\left(k_0 \sin\phi\right) = a\,\text{sinc}\left(k_0 a \sin\phi/\pi\right)$. For small angles, i.e., $\sin\phi \approx \phi$, the current amplitude falls off as $\text{sinc}\left(k_0 a \phi/\pi\right)$. Hence, the heterodyne current is at maximum when the angular separation between the signal plane wave and the reference plane wave is zero, i.e., the two plane waves are propagating exactly in the same direction. The current will go to zero when $k_0 a\phi / \pi = 1$, or $\phi = \lambda_0 / 2a$, where λ_0 is the wavelength of the laser light. To see how critical it is for the angle ϕ to be aligned in order to have any heterodyne output, we assume the size of the photodetector $2a = 1\,cm$ and the laser used is red, i.e., $\lambda_0 \approx 0.6\,\mu m$, ϕ is calculated to be about 2.3×10^{-3} degrees. Hence, to be able to work with heterodyning, one needs to have precise opto-mechanical mounts for angular rotation to minimize the angular separation of the two plane waves to be heterodyned. Figure 4.27 shows an acousto-optic heterodyne experiment, which recovers the sound frequency Ω [Cummin and Knable (1963)], where ψ_p and B have been assumed constant on the surface of the photodetector. The AOM operating in the Bragg regime, is placed in the front focal plane of the lens.

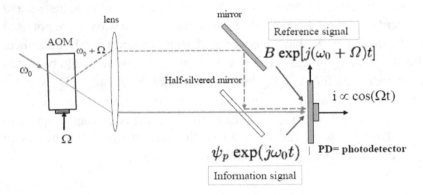

Fig. 4.27 Optical heterodyne detection.

Problems:

4.1   Verify Eq. (4.2-14).

4.2   Verify Eqs. (4.4-3) and (4.4-4).

4.3   Starting from Eq. (4.3-6), verify the Korpel-Poon equations [Eq. (4.4-6)].

4.4   Find the solution of the following infinite coupled equations:

$$\frac{da_m}{d\xi} = -j\beta a_m + \alpha(a_{m-1} - a_{m+1}),$$

where α, β, and ψ_{inc} are all real and $a_i(0) = \psi_{inc}\delta_{i0}$. Hint: Define a function $a_m(\xi) = A_m(\xi)\exp(-j\beta\xi)$.

4.5   Show that the solutions to the coupled equations given by Eqs. (4.6-3) and (4.6-4) are given by Eqs. (4.6-8a) and (4.6-8b).

4.6   For oblique plane-wave incidence on the sound column of width L, that is, $\phi_{inc} \neq 0$, find the amplitude of the m-th order diffracted light. Hint: Define a function

$$I_n = e^{jnbz} J_n\left[(a\sin bz)/b\right]$$

and show that

$$\frac{dI_n}{dz} = \frac{1}{2}ae^{2jbz} I_{n-1} - \frac{1}{2}ae^{-2jbz} I_{n+1}.$$

4.7 For *near phase-synchronous* upshifted acousto-optic Bragg diffraction, that is, $\phi_{inc} = -\phi_B(1+\delta)$, the zeroth and the first-order amplitudes evolve according to the following set of coupled equations:

$$\frac{d\psi_0}{d\xi} = -j\frac{a}{2}e^{-j\delta Q\xi/2}\psi_1$$

$$\frac{d\psi_1}{d\xi} = -j\frac{a}{2}e^{j\delta Q\xi/2}\psi_0 .$$

(a) Without solving the equations, show that $\frac{d}{d\xi}[|\psi_0|^2 + |\psi_1|^2] = 0$. Then solve for ψ_0 and ψ_1 with the boundary conditions $\psi_0(\xi = 0) = \psi_{inc}$ and $\psi_1(\xi = 0) = 0$.

4.8 As shown in Fig. P4.8, the Bragg cell operates in the Bragg regime, located in the front focal plane of the lens with focal length $F = 1\,m$. The speed of sound in the acoustic medium is $4000\,m/s$, wavelength of the light in vacuum is $\lambda_v = 0.6328\,\mu m$, and the refraction index of the medium is 1.6. $s(t)$ represents the signal feeding to the Bragg cell. Now, two test signals from radio stations WTCP and WPPB at carrier frequencies 40 MHz and 45 MHz are received by an antenna, amplified and fed to the transducer of the acousto-optic modulator. Find the following:

(a) the frequencies of the light beams on the back focal plane of the lens;

(b) the spatial separation between the two beams in the back focal plane.

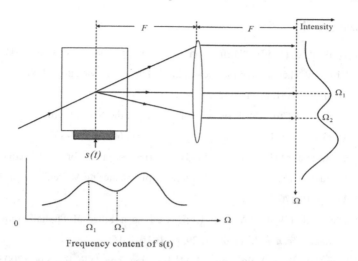

Fig. P4.8 Spectrum analyzer.

4.9 While under Bragg diffraction, the zeroth-order beam exhibits highpass spatial filtering of the incident beam. What kind of spatial filtering on the incident beam if the Bragg-scattered beam is considered? Substantiate your claim.

4.10 With reference to Fig. 4.26, find the optimum photodetector size if the two plane waves have an angular separation of 1 degree. Assume that the wavelength of light is $0.6328 \ \mu m$.

4.11 A plane wave and a spherical wave on the photodetector surface are given by the following expressions: $A_1 \exp(j\omega_0 t)$ and $A_2 \exp\left[j(\omega_0 + \Omega) t \right] \exp\left[-j \dfrac{\pi}{\lambda_0 Z}(x^2 + y^2) \right]$, respectively. The photo-detector has a uniform circular active area of radius R. Show that the heterodyne current is given by

$$i_{het}(t) \propto A_1 A_2 R^2 \, \mathrm{sinc}\left(\frac{R^2}{2\lambda_0 Z} \right) \cos\left(\Omega t - \frac{\pi R^2}{2\lambda_0 Z} \right)$$

and that the optimum photodetector size is $R_{opt} = \sqrt{\lambda_0 Z}$.

Bibliography

Aggarwal, R.R. (1950). "Diffraction of Light by Ultrasonic Waves (Deduction of Different Theories for the Generalized Theory of Raman and Nath)," *Proc. Ind. Acad. Sci.*, A31, pp. 417-426.

Alferness, R. (1976). "Analysis of Propagation at the Second-Order Bragg Angle of a Thick Holographic Grating," *J. Opt. Soc. Am.*, 66, pp. 353-362.

Appel, R. and M.G. Somekh (1993). "Series Solution for Two-Frequency Bragg Interaction Using the Korpel-Poon Multiple-Scattering Model," *J. Opt. Soc. Am. A*, 10, pp. 46-476.

Balakshy, V.I. (1984). "Acousto-Optic Cell as a Filter of Spatial Frequencies," *J. Commun. Tech. & Electronics*, 29, pp.1610-1616.

Balakshy, V.I., V.B. Voloshinov, T.M. Babkina, and D.E. Kostyuk (2005). "Optical Image Processing by Means of Acousto-Optic Spatial Filtration," *J. Modern Optics*, 52, pp. 1-20.

Banerjee, P.P. and T.-C. Poon (1991). *Principles of Applied Optics*. Irwin, Illinois.

Banerjee, P.P. and T.-C. Poon (1987). "Simulation of Bistability and Chaos in Acousto-Optic Devices," *Proc. of Midwest Symposium on Circuits & Systems,* pp. 820-823, (Syracuse, NY).

Brooks P. and C.D. Reeve (1995). "Limitations in Acousto-Optic FM Demodulators," *IEE Proc. Optoelectron.*, 142, pp. 149-156.

Cao, D., P.P. Banerjee, and T.-C. Poon (1998). "Image Edge Enhancement Using Two Cascaded Acousto-Optic Cells with Contra-Propagating Sound," *Appl. Opt.*, 37, pp. 3007-3014.

Case, S. (1979). "Fourier Processing in the Object Plane," *Opt. Lett.*, 4, pp. 286-288.

Chen, S. -T. and M. R. Chatterjee, "A Numerical Analysis and Expository Interpretation of the Diffraction of Light by Ultrasonic Waves in the Bragg and Raman-Nath Regimes Using Multiple Scattering Theory," *IEEE Trans. on Education*, 39, pp. 56-68.

Chen S. -T. and M.R. Chatterjee (1997). "Dual-Input Hybrid Acousto-Optic Set-Reset Flip-Flop and its Nonlinear Dynamics," *Appl. Opt.*, 36, pp. 3147-3154.

Cohen, M.G. and E.I. Gordon (1965). "Acoustic Beam Probing Using Optical Techniques," *Bell System Tech. Journal*, 44, pp. 693-721.

Chatterjee, M.R., T.-C. Poon and D.N. Sitter, Jr. (1990). "Transfer Function Formalism for Strong Acousto-Optic Bragg Diffraction of Light Beams with Arbitrary Profiles," *Acustica*, 71, pp. 81-92 (1990).

Chrostowski, J. and C. Delise (1982). "Bistable Optical Switching Based on Bragg Diffraction," *Opt. Commun.*, 41, pp. 71-77.

Cummins, H.Z. and N. Knable (1963). "Single Sideband Modulation of Coherent Light by Bragg Reflection from Acoustical Waves," *Proc. IEEE*, p. 1246.

Davis, J.A. and M.D. Nowak (2002). "Selective Edge Enhancement of Images with an Acousto-Optic Light Modulator," *Appl. Opt.* 41, 4835-4839.

Gabor, D. (1949). "Microscopy by Reconstructed Wavefronts," *Proc. Roy. Soc., ser. A*, 197, 454-487.

Gies, D.T and T.-C. Poon (2002). "Measurement of Acoustic Radiation Pattern in an Acousto-Optic Modulator," *Proc. IEEE SoutheastCon*, pp.441-445.

Hicks, M. and C.D. Reeve (1998). "Acousto-Optic System for Automatic Identification and Decoding of Digitally Modulated Signals," *Opt. Eng.*, 37, pp. 931-941.

IntraAction Corporation, 3719 Warren Avenue, Bellwood, IL 60104.

Klein W.R. and B.D. Cook (1967). "Unified Approach to Ultrasonic Light Diffraction," *IEEE Trans. on Sonics and Ultrasonics*, SU-14, 123-134.

Kogelnik, H. (1969). "Coupled Wave Theory for Thick Hologram Grating," *Bell Syst. Tech. J.*, 48, pp. 2909-2947.

Korpel, A. (1972). Acousto-Optics. In *Applied Solid State Science*, Vol. 3 (R. Wolfe, ed.) Academic, New York.

Korpel, A. (1988). *Acousto-Optics*. Marcel Dekker, Inc., New York and Basel.

Korpel, A., R. Adler, and B. Alpiner (1964). "Direct Observation of Optically Induced Generation and Amplification of Sound," *Applied Physics Letters*, 5, pp. 86-88.

Korpel, A. and T.-C Poon (1980). "Explicit Formalism for Acousto-Optic Multiple Plane-Wave Scattering," *J. Opt. Soc. Am.*, 70, pp. 817-820.

Magdich L.N., V.Y. Molchanov, and V. Ya (1977). "Diffraction of a Divergent Beam by Intense Acoustic Waves, " *Opt. Spectrosc.* (USSR), 42, pp. 299-302.

Marquez *et al.* (2003). "Edge-Enhanced Imaging with Polyvinyl Alcohol/Acrylamide Photopolymer Grating," *Opt. Lett.*, 28, pp. 1510-1512.

McNeil, M. and T.-C. Poon (1992). "Hybrid Acousto-Optical Bistability in the Second-Order Bragg Regime," *IEEE Proc. Southeastern Symposium on System Theory,* pp. 556-560.

Mertens, R., W. Hereman, and J.-P. Ottoy (1985). "The Raman-Nath Equations Revisited," *Proc. Ultrason. Int.,* 85, 422-428.

Phariseau, P. (1956). "On the Diffraction of Light by Progressive Supersonic Waves, " *P. Proc. Indian Acad. Sci.* 44:165.

Pieper, R.J. and T.-C. Poon (1985). "An Acousto-Optic FM Receiver Demonstrating Some Principles of Modern Signal Processing," *IEEE Trans. on Education,* Vol. E-27, No. 3, pp. 11-17.

Poon, T.-C. (2002). "Acousto-Optics," in *Encyclopedia of Physical Science and Technology,* Academic Press.

Poon, T.-C. (2005). "Heterodyning," in *Encyclopedia of Modern Optics,* Elsevier Physics, pp. 201-206.

Poon, T.-C. and S.K. Cheung (1989). "Performance of a Hybrid Bistable Device Using an Acousto-Optic Modulator," *Appl. Opt.,* (Special Issue on Spatial Light Modulators), 28, pp. 4787-4791.

Poon, T.-C. and T. Kim (2005). "Acousto-Optics with MATLAB®," *Proc. SPIE,* Vol. 5953, 59530J-1-59530J-12.

Poon, T.-C. and A. Korpel (1981a). "Feynman Diagram Approach to Acousto-Optic Scattering in the Near Bragg Region," *J. Opt. Soc. Am.,* 71, pp. 1202-1208.

Poon, T.-C. and A. Korpel (1981b). "High Efficiency Acousto-Optic Diffraction into the Second Bragg Order," *IEEE Proc. Ultrasonics Symposium,* Vol. 2, pp. 751-754.

Poon, T.-C, M.D. McNeill, and D.J. Moore (1997). "Two Modern Optical Signal Processing Experiments Demonstrating Intensity and Pulse-Width Modulation Using an Acousto-Optic Modulator," *American Journal of Physics,* 65, pp. 917-925.

Poon, T.-C. and R.J. Pieper (1983). "Construct an Optical FM Receiver," *Ham Radio,* pp. 53-56.

Raman C.V. and N.S.N. Nath (1935). "The Diffraction of Light by High Frequency Sound Waves: Part I.," *Proc. of the Indian Academy of Sciences,* 2, pp. 406-412.

Whitman R.L. and A. Korpel (1969). "Probing of Acoustic Surface Perturbations by Coherent Light," *Appl. Opt.,* 8, pp. 1567-1576.

VanderLugt, A. (1992). *Optical Signal Processing.* John Wiley & Sons, Inc., New York.

Voloshinov, V.I., V.B. Babkina, and V.Y. Molchanov (2002). "Two Dimensional Selection of Optical Spatial Frequencies by Acousto-Optic Methods," *Opt. Eng.*, 41, pp. 1273-1280.

Xia, J., D.B. Dunn, T.-C. Poon, and P.P. Banerjee (1996). "Image Edge Enhancement by Bragg Diffraction," *Opt. Commun.*,128, pp. 1-7.

Chapter 5

Electro-Optics

We have studied the effects of wave propagation through homogeneous media (Chapter 2), inhomogeneous media (Chapter 3) as well as time-varying media (Chapter 4). However, many materials (e.g., crystals) are anisotropic. In this chapter, we will study linear wave propagation in a medium that is homogeneous and isotropic magnetically (μ_0 constant) but allows for electrical anisotropy. By this we mean that the polarization produced in the medium by an applied electric field is no longer just a constant times the field, but also depends on the direction of the applied field in relation to the anisotropy of the medium. This discussion will help us understand the properties and uses of electro-optic materials for laser modulation.

5.1 The Dielectric Tensor

From Chapter 2, we know that for a linear, homogeneous, and isotropic medium, $\boldsymbol{D} = \varepsilon \boldsymbol{E}$, where ε is a scalar constant. Hence the direction of $\boldsymbol{D} = D_x \mathbf{a_x} + D_y \mathbf{a_y} + D_z \mathbf{a_z}$ is parallel to the direction of $\boldsymbol{E} = E_x \mathbf{a_x} + E_y \mathbf{a_y} + E_z \mathbf{a_z}$. For example, an applied electrical field $E_x \mathbf{a_x}$ generates a displacement $\boldsymbol{D}$ lying only the x-direction, i.e., $D_x \mathbf{a_x}$ only. This is not the case in anisotropic media. Figure 5.1 depicts a model illustrating anisotropic binding of an electron in a crystal. *Anisotropy* is taken into account by assuming different spring constants in each direction (In the isotropic case, all spring constants are equal). Hence, in general, an applied electrical field $E_x \mathbf{a_x}$ generates a $\boldsymbol{D}$ which has all the three components as follows:

$$D_x = \varepsilon_{xx} E_x, D_y = \varepsilon_{yx} E_x, D_z = \varepsilon_{zx} E_x, \quad (5.1\text{-}1a)$$

where ε_{xx}, ε_{yx}, and ε_{zx} are the corresponding permittivity components. By the same token, an applied electrical field $E_y \mathbf{a_y}$ generates

$$D_x = \varepsilon_{xy} E_y, D_y = \varepsilon_{yy} E_y, D_z = \varepsilon_{zy} E_y \quad (5.1\text{-}1b)$$

and $E_z \mathbf{a_z}$ generates

$$D_x = \varepsilon_{xz} E_z, D_y = \varepsilon_{yz} E_z, D_z = \varepsilon_{zz} E_z. \quad (5.1\text{-}1c)$$

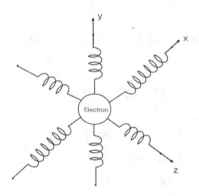

Fig. 5.1 Anisotropic binding of an electron in a crystal (the spring constants are different in each direction).

To generalize the above, if the applied electric field has all three components, i.e., $\boldsymbol{E} = E_x \mathbf{a_x} + E_y \mathbf{a_y} + E_z \mathbf{a_z}$, in place of Eq. (2.1-12a), the components of $\boldsymbol{D}$ and $\boldsymbol{E}$ are related by the following equation [Note that Eq. (2.1-12b) still holds true, as we are considering a magnetically isotropic medium]:

$$D_x = \varepsilon_{xx}E_x + \varepsilon_{xy}E_y + \varepsilon_{xz}E_z , \tag{5.1-2a}$$

$$D_y = \varepsilon_{yx}E_x + \varepsilon_{yy}E_y + \varepsilon_{yz}E_z , \tag{5.1-2b}$$

$$D_z = \varepsilon_{zx}E_x + \varepsilon_{zy}E_y + \varepsilon_{zz}E_z , \tag{5.1-2c}$$

or in the matrix form

$$\begin{pmatrix} D_x \\ D_y \\ D_z \end{pmatrix} = \begin{pmatrix} \varepsilon_{xx} & \varepsilon_{xy} & \varepsilon_{xz} \\ \varepsilon_{yx} & \varepsilon_{yy} & \varepsilon_{yz} \\ \varepsilon_{zx} & \varepsilon_{zy} & \varepsilon_{zz} \end{pmatrix} \begin{pmatrix} E_x \\ E_y \\ E_z \end{pmatrix} , \tag{5.1-3a}$$

or simply as

$$D_i = \sum_{j=1}^{3} \varepsilon_{ij} E_j , \tag{5.1-3b}$$

where $i, j = 1$ for x, 2 for y, and 3 for z. Eq. (5.1-3a) or Eq. (5.1-3b) is customarily written as

$$\boldsymbol{D} = \bar{\varepsilon}\boldsymbol{E} , \tag{5.1-4}$$

where

$$\bar{\varepsilon} = \begin{pmatrix} \varepsilon_{11} & \varepsilon_{12} & \varepsilon_{13} \\ \varepsilon_{21} & \varepsilon_{22} & \varepsilon_{23} \\ \varepsilon_{31} & \varepsilon_{32} & \varepsilon_{33} \end{pmatrix} . \tag{5.1-5}$$

The above 3 × 3 matrix is commonly known as the *dielectric tensor*. We shall now show that, in a lossless medium, the dielectric tensor has special symmetry properties. From Eqs. (2.1-3) and (2.1-4), two of the Maxwell's equations in the medium without a current are

$$\nabla \times \boldsymbol{E} = -\frac{\partial \boldsymbol{B}}{\partial t} , \tag{5.1-6}$$

and

$$\nabla \times \mathcal{H} = \frac{\partial \mathcal{D}}{\partial t} \ , \tag{5.1-7}$$

where $\mathcal{B} = \mu \mathcal{H}$ with $\mu = \mu_0$ for magnetically isotropic medium. Now, taking the dot product of $\mathcal{E}$ with Eq. (5.1-7), we have

$$\mathcal{E} \cdot \left(\nabla \times \mathcal{H} \right) = \mathcal{E} \cdot \frac{\partial \mathcal{D}}{\partial t} \ .$$

Since $\nabla \cdot \left(\mathcal{E} \times \mathcal{H} \right) = \mathcal{H} \cdot \left(\nabla \times \mathcal{E} \right) - \mathcal{E} \cdot \left(\nabla \times \mathcal{H} \right)$ is a vector identity, we can write the above equation as

$$\mathcal{H} \cdot \left(\nabla \times \mathcal{E} \right) - \nabla \cdot \left(\mathcal{E} \times \mathcal{H} \right) = \mathcal{E} \cdot \frac{\partial \mathcal{D}}{\partial t} \ . \tag{5.1-8}$$

Now using Eq. (5.1-6), we have

$$\mathcal{H} \cdot \left(\nabla \times \mathcal{E} \right) = \mathcal{H} \cdot \left(-\frac{\partial \mathcal{B}}{\partial t} \right) . \tag{5.1-9}$$

Substituting Eq. (5.1-9) into (5.1-8), we write

$$-\nabla \cdot \left(\mathcal{E} \times \mathcal{H} \right) = \mathcal{H} \cdot \left(\frac{\partial \mathcal{B}}{\partial t} \right) + \mathcal{E} \cdot \frac{\partial \mathcal{D}}{\partial t} \ . \tag{5.1-10}$$

From the definition of the Poynting vector $S = \mathcal{E} \times \mathcal{H}$ [see Eq. (2.2-32)], we can re-write Eq. (5.1-10) as

$$-\nabla \cdot S = \mathcal{H} \cdot \left(\frac{\partial \mathcal{B}}{\partial t} \right) + \mathcal{E} \cdot \frac{\partial \mathcal{D}}{\partial t}. \tag{5.1-11}$$

But the *continuity equation for energy flow* reads

$$\nabla \cdot S + \frac{dW}{dt} = 0 , \tag{5.1-12}$$

which states that the time rate of change of electromagnetic energy density W is equal to the outflow of the energy flux $-\nabla \cdot S$. Because

$$W = \frac{1}{2}\boldsymbol{E} \cdot \boldsymbol{D} + \frac{1}{2}\boldsymbol{\mathcal{H}} \cdot \boldsymbol{B}, \tag{5.1-13}$$

we then have

$$\frac{dW}{dt} = \frac{d}{dt}\left\{ \frac{1}{2}\sum_{i,j=1}^{3} E_i \varepsilon_{ij} E_j \right\} + \frac{1}{2}\left\{ \frac{d\mathcal{H}}{dt} \cdot (\mu \mathcal{H}) + \mathcal{H} \cdot \frac{d\mu \mathcal{H}}{dt} \right\}$$

$$= \frac{1}{2}\left\{ \sum_{i,j=1}^{3} \frac{dE_i}{dt}\varepsilon_{ij} E_j + \sum_{i,j=1}^{3} E_i \varepsilon_{ij} \frac{dE_j}{dt} \right\} + (\mu \mathcal{H}) \cdot \frac{d\mathcal{H}}{dt}. \tag{5.1-14}$$

By using Eqs. (5.1-11), (5.1-12) and (5.1-14), we have

$$\frac{1}{2}\left\{ \sum_{i,j=1}^{3} \frac{dE_i}{dt}\varepsilon_{ij} E_j + \sum_{i,j=1}^{3} E_i \varepsilon_{ij} \frac{dE_j}{dt} \right\} = \boldsymbol{E} \cdot \frac{\partial \boldsymbol{D}}{\partial t}$$

$$= \sum_{i=1}^{3} E_i \frac{d}{dt}\left(\sum_{j=1}^{3} \varepsilon_{ij} E_j \right)$$

$$= \sum_{i,j=1}^{3} E_i \varepsilon_{ij} \frac{dE_j}{dt}$$

Therefore, from above, we have

$$\frac{1}{2}\sum_{i,j=1}^{3} \frac{dE_i}{dt}\varepsilon_{ij} E_j = \frac{1}{2}\sum_{i,j=1}^{3} E_i \varepsilon_{ij} \frac{dE_j}{dt} \tag{5.1-15}$$

Now, exchanging the index i and j on the right-hand side of Eq. (5.1-15) does not change its value. Eq. (5.1-15) then becomes

$$\sum_{i,j=1}^{3} \frac{dE_i}{dt}\varepsilon_{ij} E_j = \sum_{i,j=1}^{3} E_j \varepsilon_{ji} \frac{dE_i}{dt}$$

which means that

$$\varepsilon_{ij} = \varepsilon_{ji} \tag{5.1-16}$$

and we have just shown that the dielectric tensor is *symmetric* in a lossless medium. For a lossy medium, ε_{ij} is complex, which leads to complex refraction indices resulting in absorption by the medium. Restricting ourselves to lossless media, we see from Eq. (5.1-16) that ε_{ij} has only six independent elements. Now, it is well known that any real symmetric matrix can be diagonalized through a coordinate transformation. Hence, the dielectric tensor can assume the diagonal form

$$\overline{\varepsilon} = \begin{pmatrix} \varepsilon_x & 0 & 0 \\ 0 & \varepsilon_y & 0 \\ 0 & 0 & \varepsilon_z \end{pmatrix}. \tag{5.1-17}$$

The new coordinate system is called the *principal axis system*, the three dielectric permittivities ε_x, ε_y and ε_z are known as the *principal dielectric constants*, and the Cartesian coordinate axes are called the *principal axes*. Note that the principal axis directions are special directions in the crystal such that an $\mathcal{E}$ applied along any one of them generates $\mathcal{D}$ parallel to it. This is evident as $\mathcal{D} = \overline{\varepsilon}\mathcal{E}$ with $\overline{\varepsilon}$ represented by Eq. (5.1-17). Three crystal classes (shown in Table 5.1) can be identified in terms of Eq. (5.1-17): *cubic*, *uniaxial*, and *biaxial*. Because most of the crystals used for electro-optic devices are uniaxial, we will therefore concentrate only on these types of crystals in our subsequent discussions. Note that for uniaxial crystals, the axis characterized by ε_z, is called the *optic axis*. When $\varepsilon_z > \varepsilon_x = \varepsilon_y$, the crystal is *positive uniaxial*, and when $\varepsilon_z < \varepsilon_x = \varepsilon_y$, it is *negative uniaxial*.

Table 5.1 Crystal classes and some common examples.

	Cubic	Uniaxial	Biaxial
Principal axis system	$\begin{pmatrix} \varepsilon & 0 & 0 \\ 0 & \varepsilon & 0 \\ 0 & 0 & \varepsilon \end{pmatrix}$	$\begin{pmatrix} \varepsilon_x & 0 & 0 \\ 0 & \varepsilon_x & 0 \\ 0 & 0 & \varepsilon_z \end{pmatrix}$	$\begin{pmatrix} \varepsilon_x & 0 & 0 \\ 0 & \varepsilon_y & 0 \\ 0 & 0 & \varepsilon_z \end{pmatrix}$
Common examples	Sodium chloride Diamond	Quartz (positive, $\varepsilon_x < \varepsilon_z$) Calcite (negative, $\varepsilon_x > \varepsilon_z$)	Mica Topaz

5.2   Plane-Wave Propagation in Uniaxial Crystals: Birefringence

In order to advance a consolidated treatment of plane-wave propagation in uniaxial crystals, we first study the constraint imposed on $\mathcal{E}$ and $\mathcal{D}$ by Maxwell's equations and then introduce the constitutive relation between $\mathcal{E}$ and $\mathcal{D}$ to eventually find an equation in $\mathcal{E}$, which will then be used to investigate plane wave propagation in the crystals.

We assume that the electromagnetic wave is a plane wave interacting within the crystal, i.e., all the dependent variables in Maxwell's equations, namely $\mathcal{E}, \mathcal{D}, \mathcal{B}$ and $\mathcal{H}$ vary according to $\exp[j(\omega_0 t - \mathbf{k_0} \cdot \mathbf{R})]$ and have constant amplitudes. In such a time-varying field, the operations d/dt and ∇ in the Maxwell's equations may be replaced according to the following rules:

$$\frac{d}{dt} \to j\omega_0 \tag{5.2-1a}$$

$$\nabla \to -j\mathbf{k_0} = -jk_0\mathbf{a_k} . \tag{5.2-1b}$$

Equations (5.1-6) and (5.1-7) then become

$$\mathbf{k_0} \times \mathcal{E} = k_0\mathbf{a_k} \times \mathcal{E} = \omega_0\mu\mathcal{H} \tag{5.2-2a}$$

and

$$\mathbf{k}_0 \times \mathcal{H} = k_0 \mathbf{a}_{\mathbf{k}} \times \mathcal{H} = -\omega_0 \mathcal{D}, \qquad (5.2\text{-}2b)$$

respectively, where we have used the fact that $\mathcal{B} = \mu_0 \mathcal{H}$ and $\mathbf{a}_{\mathbf{k}} = \mathbf{k}_0 / |\mathbf{k}_0| = \mathbf{k}_0 / k_0$. Now, eliminating $\mathcal{H}$ between Eqs. (5.2-2a) and (5.2-2b), we obtain

$$\mathcal{D} = \frac{k_0^{\,2}}{\omega_0^{\,2} \mu_0} [\mathcal{E} - (\mathbf{a}_{\mathbf{k}} \cdot \mathcal{E})\mathbf{a}_{\mathbf{k}}]. \qquad (5.2\text{-}3)$$

For electrically isotropic media, $\mathbf{a}_{\mathbf{k}} \cdot \mathcal{E} = 0$, i.e., there is no component of the electric field along the direction of propagation (see Chapter 3), Equation (5.2-3) reduces to

$$\mathcal{D} = \frac{k_0^{\,2}}{\omega_0^{\,2} \mu_0} \mathcal{E} = \frac{1}{u^2 \mu_0} \mathcal{E} = \varepsilon \mathcal{E}. \qquad (5.2\text{-}4)$$

However, in anisotropic crystals, $\mathcal{D} = \bar{\varepsilon} \mathcal{E}$, [see Eq. (5.1-4)]; hence Equation (5.2-3) becomes

$$\bar{\varepsilon} \mathcal{E} = \frac{k_0^{\,2}}{\omega_0^{\,2} \mu_0} [\mathcal{E} - (\mathbf{a}_{\mathbf{k}} \cdot \mathcal{E})\mathbf{a}_{\mathbf{k}}]. \qquad (5.2\text{-}5)$$

This equation is the starting equation for the investigation of plane wave propagation in anisotropic media.

Before we illustrate the use of Eq. (5.2-5) in some examples, we make some comments on the relative directions of the vector fields. From Eq. (5.2-2), we see that $\mathcal{H}$ or $\mathcal{B}$ is at right angles to $\mathbf{k}_0$, $\mathcal{E}$ and $\mathcal{D}$, which are co-planar. Furthermore, $\mathcal{D}$ is orthogonal to $\mathbf{k}_0$ as suggested by Eq. (2.1-1) as $\nabla \cdot \mathcal{D} = -j \mathbf{k}_0 \cdot \mathcal{D} = 0$ for source-free region. This means that for a general anisotropic medium, $\mathcal{D}$ is perpendicular to the direction of propagation, even though E may not be so because of the anisotropy. This is summarized in Fig. 5.2.

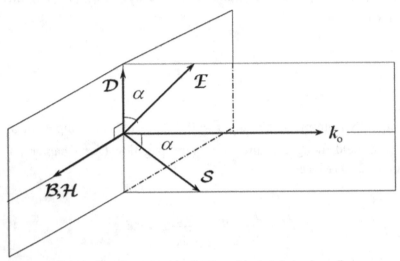

Fig. 5.2 Directions of various field quantities in anisotropic media.

Note that the Poynting vector $S = E \times H$ (which determines the direction of energy flow) is different from the direction of propagation of the wavefronts denoted by $\mathbf{k_0}$ and that the angle α between E and D is the same as the angle between $\mathbf{k_0}$ and S. In fact, we can distinguish between the phase velocity $u_p = \omega_0 / k_0$ and the group velocity $\boldsymbol{u_g}$, which is defined as

$$u_g = S / W, \qquad\qquad (5.2\text{-}6)$$

where W has been defined in Eq. (5.1-13). The group velocity is in the direction of S, and we can show that the phase velocity is the projection of the group velocity onto the direction $\mathbf{k_0}$ given by

$$u_p = |\mathbf{u}_g| \cos\alpha . \tag{5.2-7}$$

We will now use Eq. (5.2-5) to analyze plane-wave propagation in uniaxial crystals. Suppose that a plane wave is polarized so that its electric field $\boldsymbol{E}_a$ is perpendicular to the optic axis, and the propagation vector is at an angle θ with respect to the optic axis as shown in Fig. 5.3. The x direction is pointing normally inward into the paper. Since $\boldsymbol{E}$ varies according to $\exp[j(\omega_0 t)]$, we can hence write $\boldsymbol{E}_a = E_a \mathbf{a_x} \exp[j(\omega_0 t)]$. Substituting into Eq. (5.2-5) and having $\varepsilon_x = \varepsilon_y$ we have

$$\begin{pmatrix} \varepsilon_x & 0 & 0 \\ 0 & \varepsilon_y & 0 \\ 0 & 0 & \varepsilon_z \end{pmatrix} \begin{pmatrix} E_a\mathbf{a_x} \\ 0 \\ 0 \end{pmatrix} = \frac{k_0^2}{\omega_0^2 \mu_0} [E_a\mathbf{a_x} - (\mathbf{a_k} \cdot E_a\mathbf{a_x})\mathbf{a_k}]. \tag{5.2-8}$$

Since $E_a\mathbf{a_x}$ is along the direction of one of the principal axes, its $\boldsymbol{D}$ field is therefore parallel to it. Hence the term $\mathbf{a_k} \cdot E_a\mathbf{a_x}$ in the above equation is zero as $\mathbf{k}_0 \cdot \boldsymbol{D} = 0$. Finally, by comparing the x-component of Eq. (5.2-8), we have

$$\varepsilon_x E_a = \frac{k_0^2}{\omega_0^2 \mu_0} E_a$$

or

$$k_0^2 = \omega_0^2 \mu_0 \varepsilon_x = \omega_0^2 \mu_0 \varepsilon_y. \tag{5.2-9}$$

Hence the phase velocity corresponding to the wave polarized along the x-direction is

$$u_x = u_1 = \frac{\omega_0}{k_0} = \frac{1}{\sqrt{\mu_0 \varepsilon_x}} = \frac{1}{\sqrt{\mu_0 \varepsilon_y}} = u_2 . \qquad (5.2\text{-}10)$$

We can show that for any $\boldsymbol{\mathcal{E}}$ that has no component along the optic axis, the same velocity (u_1 or u_2) is obtained.

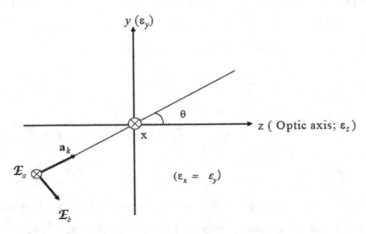

Fig. 5.3 Plane-wave propagation in a uniaxial crystal.

Consider a second example, where $\boldsymbol{\mathcal{E}}_b$ has a polarization vector as $E_{by}\mathbf{a_y} + E_{bz}\mathbf{a_z}$ as shown in Fig. 5.3. Equation (5.2-5) then reduces to the following two component equations

$$\omega_0^2 \mu_0 \varepsilon_y E_{by} = k_0^2 [E_{by} - (\sin\theta\, E_{by} + \cos\theta\, E_{bz})\sin\theta], \qquad (5.2\text{-}11a)$$

and

$$\omega_0^2 \mu_0 \varepsilon_z E_{bz} = k_0^2 [E_{bz} - (\sin\theta\, E_{by} + \cos\theta\, E_{bz})\cos\theta], \qquad (5.2\text{-}11b)$$

where $\mathbf{a_k}$ in Eq. (5.2-5) has been written as $\mathbf{a_k} = \sin\theta\,\mathbf{a_y} + \cos\theta\,\mathbf{a_z}$. We can re-express Eqs. (5.2-11a) and (5.2-11b) as

$$\begin{pmatrix} \omega_0^2 \mu_0 \varepsilon_y - k_0^2 \cos^2 \theta & k_0^2 \cos \theta \sin \theta \\ k_0^2 \cos \theta \sin \theta & \omega_0^2 \mu_0 \varepsilon_z - k_0^2 \sin^2 \theta \end{pmatrix} \begin{pmatrix} E_{by} \\ E_{bz} \end{pmatrix} = 0 . \quad (5.2\text{-}12)$$

For nontrivial solutions, the determinant of the 2 × 2 matrix should be zero, and this gives

$$k_0^2 = \frac{k_{0y}^2 k_{0z}^2}{k_{0y}^2 \sin^2 \theta + k_{0z}^2 \cos^2 \theta}, \quad (5.2\text{-}13)$$

where $k_{0y}^2 = \omega_0^2 \mu_0 \varepsilon_y$ and $k_{0z}^2 = \omega_0^2 \mu_0 \varepsilon_z$. The total phase velocity is accordingly given by

$$u = \frac{\omega_0}{k_0} = \sqrt{(u_2 \cos \theta)^2 + (u_3 \sin \theta)^2} , \quad (5.2\text{-}14)$$

where $u_2 = 1/\sqrt{\mu_0 \varepsilon_y}$ and $u_3 = 1/\sqrt{\mu_0 \varepsilon_z}$. Note that when $\theta = 0$, the wave propagates along the optic axis with the electric field (in the $-\mathbf{a}_y$ direction) perpendicular to it. The phase velocity is $u_2 = 1/\sqrt{\mu_0 \varepsilon_y}$. In fact, the phase velocity of the wave is the same for any wave that has no polarization vector component along the optic axis, that is $u_1 = u_2 = 1/\sqrt{\mu_0 \varepsilon_x} = c/n_0$ as $\varepsilon_x = \varepsilon_y$ in uniaxial crystals, where $n_0 = \sqrt{\varepsilon_x / \varepsilon_0}$ is the *ordinary refractive index*. The wave associated with the two identical refractive indices is often called the *ordinary wave* or simply as the *o-ray*. Now, when $\theta = \pi/2$, the polarization vector lies along the optic axis. The phase velocity of the wave is $u_3 = 1/\sqrt{\mu_0 \varepsilon_z} = c/n_e$, where $n_e = \sqrt{\varepsilon_z / \varepsilon_0}$ is the *extraordinary refractive index* and the wave associated with the dissimilar refractive index is the *extraordinary wave* or *e-ray*. This phenomenon, in which the phase velocity of an optical wave propagating in the crystal depends on the direction of its polarization, is called *birefringence*. We can show that when a light beam is incident on a uniaxial crystal in a direction other than the optic axis, two beams will emerge from the crystal, due to the two dissimilar indices of refraction n_e and n_0, and both are polarized perpendicular to each other. Therefore, if we view an object through the crystal, two images will be observed. This remarkable optical phenomenon in crystals is called *double refraction*. In what follows, we analyze a situation that a wave of random polarization is incident normally on a uniaxial crystal as shown in Fig. 5.4.

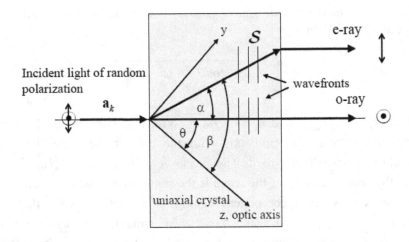

Fig. 5.4 Double refraction in a uniaxial crystal.

Note: ᵒdenotes direction of polarization pointing into the paper, where⬍ denotes pointing up and down on the plane of the paper.

The direction of the incident wave is denoted by $\mathbf{a_k}$ and the wave makes an angle θ with the z-axis, the optic axis. We will find the angle β between the e-ray and the optic axis. Let us find the direction of energy flow for the e-ray, which is done by evaluating the Poynting vector S. Hence, we have

$$S = \mathcal{E} \times \mathcal{H}$$

$$= \frac{k_0}{\omega_0 \mu} \mathcal{E} \times (\mathbf{a_k} \times \mathcal{E}),$$

where we have used Eq. (5.2-2a). By using vector identity for the two cross products, we can re-write the above equation to get

$$S = \frac{k_0}{\omega_0 \mu} [\mathbf{a_k} |\mathcal{E}|^2 - \mathcal{E}(\mathcal{E} \cdot \mathbf{a_k})] \quad . \tag{5.2-15}$$

For a wave where $\mathcal{E}$ is normal to the optic axis, identical to the situation where $\mathcal{E}_a$ is shown in Fig. 5.3 we have $\mathcal{E} \cdot \mathbf{a_k} = 0$ and hence, S is parallel to $\mathbf{a_k}$. This wave, the o-ray, passes straight through the crystal without deflection. However, if the wave where $\mathcal{E}$ lies on the $y-z$ plane, identical to the case where $\mathcal{E}_b$, is shown in Fig. 5.3, this wave, i.e., the e-ray does not pass straight through the crystal and will be refracted. We note that the wavefronts are propagating along the $\mathbf{a_k}$ direction but the direction of the energy will be along the Poynting vector. We analyze the situation as follows.

Again, assuming $\mathcal{E}$ has a polarization of the form $E_y \mathbf{a_y} + E_z \mathbf{a_z}$, which is on the $y-z$ plane, and also we recognize that $\mathbf{a_k} = \sin\theta \, \mathbf{a_y} + \cos\theta \, \mathbf{a_z}$. By taking E_y and E_z to be real for simplicity, we can write Eq. (5.2-15) as

$$S = \frac{k_0}{\omega_0 \mu} [(E_y^2 + E_z^2)\mathbf{a_k} - (E_y \mathbf{a_y} + E_z \mathbf{a_z})(E_y \sin\theta + E_z \cos\theta)]. \tag{5.2-16}$$

Now, the angle β as shown in Fig. 5.4 can be determined by

$$\tan\beta = \frac{S_y}{S_z} = \frac{(E_y^2 + E_z^2)\sin\theta - E_y(E_y \sin\theta + E_z \cos\theta)}{(E_y^2 + E_z^2)\cos\theta - E_z(E_y \sin\theta + E_z \cos\theta)}, \tag{5.2-17}$$

where S_y and S_z are the y- and z-components of S, respectively. Equation (5.2-17) can be simplified to the following form

$$\tan \beta = \frac{(E_z / E_y)\sin\theta - \cos\theta}{(E_y / E_z)\cos\theta - \sin\theta}.$$ (5.2-18)

From the relations given by Eqs. (5.2-12) and (5.2-13), we can find the ratio of E_z and E_y. By incorporating the ratio into Eq. (5.2-18), we can finally obtain the following equation:

$$\tan \beta = \left(\frac{k_{0y}}{k_{0z}}\right)^2 \tan\theta = \left(\frac{n_0}{n_e}\right)^2 \tan\theta.$$ (5.2-19)

Considering Potassium Dihydrogen Phosphate (KDP) for which $n_0 = 1.50737$ and $n_e = 1.46685$, and for incident angle at $45°$ with respect to the optic axis, i.e., $\theta = 45°$, we use Eq. (5.2-19) to calculate the angle between the o-ray and the e-ray, i.e., $\alpha = \beta - \theta$, to be about $1.56°$. For $\theta = 0°$ or $90°$, the angle between the o-ray and the e-ray is zero, implying that both the rays propagate along the same direction, which is consistent with the previous result.

5.3 Applications of Birefringence: Wave Plates

Consider a plate of thickness d made of a uniaxial material. The optic axis is along the z direction, as shown in Fig. 5.5.

Let a linearly polarized optical field incident on crystal at $x = 0$ cause a field in the crystal at $x = 0^+$ of the form

$$\mathbf{E}_{inc} = \mathrm{Re}[E_0(\mathbf{a_y} + \mathbf{a_z})e^{j\omega_0 t}].$$ (5.3-1)

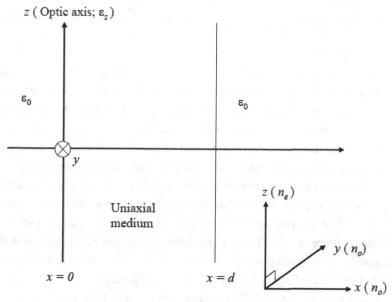

Fig. 5.5 A wave plate.

The wave travels through the plate, and at the right edge of the plate $(x = d)$ the field can be represented as

$$\boldsymbol{\mathcal{E}}_{out} = \text{Re}\left[E_0 \left[\exp\left(-j\frac{\omega_0}{u_2}d \right)\mathbf{a_y} + \exp\left(-j\frac{\omega_0}{u_3}d \right)\mathbf{a_z} \right] e^{j\omega_0 t} \right] \quad (5.3\text{-}2)$$

$$= \text{Re}\left[E_0 \exp\left(-j\frac{\omega_0}{u_2}d \right)(\mathbf{a_y} + \exp(-j\Delta\phi)\mathbf{a_z})e^{j\omega_0 t} \right]$$

The two plane-polarized waves are the ordinary wave (along $\mathbf{a_y}$) and the extraordinary wave (along $\mathbf{a_z}$) and they acquire a different phase as they propagate through the crystal. The relative phase shift $\Delta\phi$ between the two waves is

$$\Delta\phi = \left(\frac{\omega_0}{u_3} - \frac{\omega_0}{u_2} \right) d = \frac{\omega_0}{c}(n_e - n_o)d = \frac{2\pi}{\lambda_v}(n_e - n_o)d, \qquad (5.3\text{-}3)$$

where λ_v is the wavelength of light in vacuum. If $n_e > n_o$, the extraordinary wave lags the ordinary wave in phase, that is, the ordinary wave travels faster, whereas if $n_e < n_o$, the opposite is true. Such a phase shifter is often referred to as a *compensator* or a *retardation plate*. The directions of polarization for the two allowed waves are mutually orthogonal and are usually called the *slow* and *fast axes* of the crystal. If $n_e > n_o$, the z-axis is the slow axis and the y-axis is the fast axis. Since the plate surfaces are perpendicular to the principal x axis, the direction in which the wave propagates, the crystal plate is then called *x-cut*. If $\Delta\phi = 2m\pi$, where m is an integer (by introducing the right thickness d), the compensator is called a *full-wave plate*. The extraordinary wave is retarded or advanced by a full cycle of λ_v with respect to the ordinary wave. The emerging beam will have the same state of polarization as the incident beam. If $\Delta\phi = (2m+1)\pi$ and $(m+1/2)\pi$, we have a *half-wave plate* and a *quarter-wave plate*, respectively. A half-wave plate can rotate the plane of polarization of the incident wave. This is illustrated in Fig.5.6.

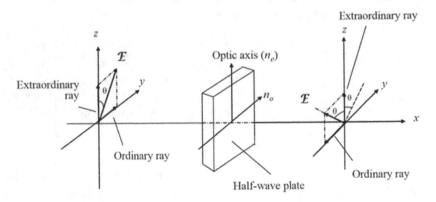

Fig. 5.6   Rotation of polarization by a half-wave plate.

Note that a relative phase change of π is equivalent to reversing one of the components of $\boldsymbol{E}$. The combination of the two components at the exit of the plate shows that the polarization direction has been rotated through twice the angle θ. It should also be noted that the rotation is independent of which component is reversed, as the polarization direction only specifies the line along which the vector $\mathcal{E}$ oscillates. Although a halfwave plate can rotate the polarization direction, a quarter-wave plate can transform a plane-polarized wave to a circularly polarized wave, as the plate introduces a relative phase shift of $\pi/2$ between the extraordinary and ordinary waves. Finally, it should be noted that for any other value of $\Delta\phi$, the emerging wave will be elliptically polarized for a plane-polarized incident wave.

5.4 The Index Ellipsoid

We have shown mathematically that for a plane-polarized wave propagating in any given direction in a uniaxial crystal, there are *two allowed polarizations*, one along the optic axis and the other perpendicular to it. As shown in Section 5.2, the total phase velocity of a wave propagating in an arbitrary direction depends on the velocities of waves polarized solely along the directions of the principal axes and on the direction of propagation, θ, of the wave. A convenient method to figure out the directions of polarization of the two allowed waves and their phase velocities is through the *index ellipsoid*, a mathematical entity written as

$$\frac{x^2}{n_x^2} + \frac{y^2}{n_y^2} + \frac{z^2}{n_z^2} = 1, \qquad (5.4\text{-}1)$$

where $n_x^2 = \varepsilon_x / \varepsilon_0$, $n_y^2 = \varepsilon_y / \varepsilon_0$ and $n_z^2 = \varepsilon_z / \varepsilon_0$. Figure 5.7 shows the index ellipsoid.

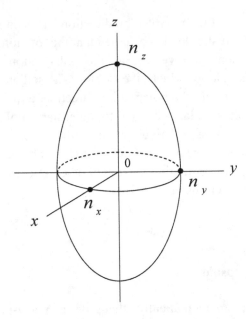

Fig. 5.7. The index ellipsoid described by Eq. (5.4-1).

Indeed we can derive Eq. (5.4-1) by recognizing that the electric part of

$$W_e = \frac{1}{2}\boldsymbol{E}\cdot\boldsymbol{D}$$

or the energy density of the EM field is

$$2W_e = (E_x\mathbf{a}_x \quad E_y\mathbf{a}_y \quad E_z\mathbf{a}_z)\begin{pmatrix} \varepsilon_x & 0 & 0 \\ 0 & \varepsilon_y & 0 \\ 0 & 0 & \varepsilon_z \end{pmatrix}\begin{pmatrix} E_x\mathbf{a}_x \\ E_y\mathbf{a}_y \\ E_z\mathbf{a}_z \end{pmatrix}$$

$$= \varepsilon_x E_x^2 + \varepsilon_y E_y^2 + \varepsilon_z E_z^2$$

$$= \frac{D_x^2}{\varepsilon_x} + \frac{D_y^2}{\varepsilon_y} + \frac{D_z^2}{\varepsilon_z} \tag{5.4-2}$$

as $D_i = \varepsilon_i E_i$ where $i = x, y$ or z. If we write x, y, and z in place of $\sqrt{D_i / 2W_e\varepsilon_0}$, and consider these as Cartesian coordinates in space, Equation (5.4-2) becomes

$$1 = \frac{x^2}{\varepsilon_x / \varepsilon_0} + \frac{y^2}{\varepsilon_y / \varepsilon_0} + \frac{z^2}{\varepsilon_z / \varepsilon_0}, \qquad (5.4\text{-}3)$$

which is identical to the index ellipsoid given by Eq. (5.4-1). In general, any point on the surface of the index ellipsoid is a solution of the index or the field D because $x, y,$ and z are proportional to D_i in order to obtain Eq. (5.4-3) from Eq. (5.4-2). Therefore, the $x, y,$ and z axes in the index ellipsoid coincides in directions with the principal axes. Indeed from Eq. (5.4-1), we can determine the respective refractive indices as well as the two allowed polarizations of D for a given direction of propagation in crystals. Let us see how this can be done. For a given direction of propagation denoted by $\mathbf{k}_0$, the field D associated with $\mathbf{k}_0$ must lie in a plane perpendicular to $\mathbf{k}_0$ because $\nabla \cdot D = -j\mathbf{k}_0 \cdot D = 0$ for source-free region. The plane intersects the ellipsoid in the ellipse A, and the directions of the major and minor axes correspond to the two orthogonal directions of the displacements D_1 and D_2 , which are the orthogonal decomposition of D . This is illustrated in Fig. 5.8, which has been drawn for a uniaxial crystal. The lengths of the major and minor axes give the respective refractive indices and we can thereafter find the phase velocities of D_1 and D_2 .

As an example, we consider a wave propagating with $\mathbf{k}_0$ along the y-direction, i.e., $\theta = \pi / 2$. To determine the phase velocities of the two allowed waves, we take a section of the ellipsoid perpendicular to $\mathbf{k}_0$ and the ellipse is given by the following equation

$$\frac{x^2}{n_o^2} + \frac{z^2}{n_e^2} = 1.$$

Hence, we will have two waves $\mathcal{D}_1$ and $\mathcal{D}_2$ polarized along x- and z-directions with phase velocities $u_1 = c/n_0$ and $u_3 = c/n_e$, respectively. $\mathcal{D}_1$ is, therefore, the ordinary wave and $\mathcal{D}_2$ is the extraordinary wave.

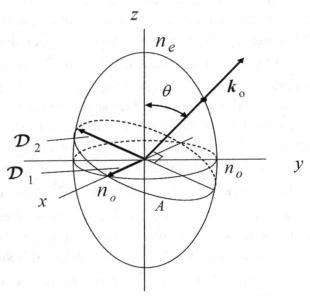

Fig. 5.8 Two allowed polarization $\mathcal{D}_1$ and $\mathcal{D}_2$.

If θ is arbitrary but on the $y-z$ plane, the index of refraction $n_e(\theta)$ along $\mathcal{D}_2$ can be determined using Fig. 5.9. Employing the relations

$$n_e^2(\theta) = z^2 + y^2,$$
(5.4-4a)

$$\frac{z}{n_e(\theta)} = \sin\theta,$$
(5.4-4b)

and the equation of the ellipse

$$\frac{y^2}{n_o^2} + \frac{z^2}{n_e^2} = 1,$$
(5.4-4c)

we have

$$\frac{1}{n_e^2(\theta)} = \frac{\cos^2\theta}{n_o^2} + \frac{\sin^2\theta}{n_e^2}.$$
(5.4-5)

Note that Eq. (5.4-5) can be obtained directly from Eq. (5.2-14). Also note that from Fig. 5.9, we can observe immediately that when $\theta = 0$, that is, the wave is propagating along the optic axis, no birefringence is observed as the amount of birefringence is zero, i.e., $n_e(0) - n_0 = 0$. Also, the amount of birefringence, $n_e(0) - n_0$, depends on the propagation direction and it is maximum at the value of $n_e - n_o$ when the propagation direction is perpendicular to the optic axis, $\theta = 90°$.

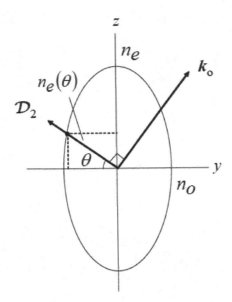

Fig. 5.9 Cross-section of index ellipsoid illustrating the value of refractive index depending on the direction of wave propagation.

5.5 Electro-Optic Effect in Uniaxial Crystals

Having been introduced to wave propagation in anisotropic media, we are now in a position to analyze the electro-optic effect, which is inherently anisotropic. As seen later, we can effectively study this using the index ellipsoid concepts. In the following Section, we will study the applications of the electro-optic effect in areas discussed previously in connection with acousto-optics, for example, intensity and phase modulation.

The *electro-optic effect* is, loosely speaking, the change in n_e and n_o that is caused by an applied electric field. In this chapter, we will only discuss the *linear* (or *Pockels-type*) *electro-optic effect*, meaning the case where the change in n_e *and* n_o is linearly proportional to the applied field. The other case, namely, where n_e and n_o depend nonlinearity on the applied field, is called the *Kerr effect* and has been discussed in the context of nonlinear optics in Chapter 3.

Mathematically, the electro-optic effect can be best represented as a deformation of the index ellipsoid due to an external electric field. Instead of specifying the changes in the refractive indices, it is more convenient to describe the changes in $1/n^2$, i.e., $\Delta(1/n^2)$, due to the external electric field. Restricting our analysis to the linear electro-optic (Pockels) effects and uniaxial crystals, the general expression for the ellipsoid becomes

$$\left[\frac{1}{n_o^2} + \Delta\left(\frac{1}{n^2}\right)_1\right]x^2 + \left[\frac{1}{n_o^2} + \Delta\left(\frac{1}{n^2}\right)_2\right]y^2 + \left[\frac{1}{n_e^2} + \Delta\left(\frac{1}{n^2}\right)_3\right]z^2$$

$$+2\Delta\left(\frac{1}{n^2}\right)_4 yz + 2\Delta\left(\frac{1}{n^2}\right)_5 xz + 2\Delta\left(\frac{1}{n^2}\right)_6 xy = 1, \qquad (5.5\text{-}1)$$

where

$$\Delta\left(\frac{1}{n^2}\right)_i = \sum_{j=1}^{3} r_{ij} E_j, \quad i = 1,\ldots, 6,$$

and where r_{ij} are called linear electro-optic (or Pockels) coefficients. The E_js are the components of the externally applied electric field in the x, y, and z directions. Note that when the applied electric field is zero, i.e., $E_j = 0$, Eq. (5.5-1) reduces to Eq. (5.4-1), where $n_x = n_o = n_y$, $n_z = n_e$. We can express $\Delta\left(\frac{1}{n}\right)_i$ in matrix form as

$$
\begin{pmatrix}
\Delta\left(\dfrac{1}{n^2}\right)_1 \\[1.2em]
\Delta\left(\dfrac{1}{n^2}\right)_2 \\[1.2em]
\Delta\left(\dfrac{1}{n^2}\right)_3 \\[1.2em]
\Delta\left(\dfrac{1}{n^2}\right)_4 \\[1.2em]
\Delta\left(\dfrac{1}{n^2}\right)_5 \\[1.2em]
\Delta\left(\dfrac{1}{n^2}\right)_6
\end{pmatrix}
=
\begin{pmatrix}
r_{11} & r_{12} & r_{13} \\
r_{21} & r_{22} & r_{23} \\
r_{31} & r_{32} & r_{33} \\
r_{41} & r_{42} & r_{43} \\
r_{51} & r_{52} & r_{53} \\
r_{61} & r_{62} & r_{63}
\end{pmatrix}
\begin{pmatrix}
E_1 \\
E_2 \\
E_3
\end{pmatrix}.
\qquad (5.5\text{-}2)
$$

The 6×3 matrix is often written as $[r_{ij}]$ and known as the *linear electro-optic tensor*. The tensor contains 18 elements and they are necessary in the most general case, when no symmetry is present in the crystal. But many of the 18 elements are zero and some of the non-zero elements have the same value, depending on the symmetry of the crystal. Table 5.2 lists the non-zero values of some of the representative crystals. Using Eq. (5.5-1) and Table 5.2, we can find the equation of the index ellipsoid in the presence of an external applied field.

Table 5.2 Pockels coefficients.

Material	$r_{ij}(10^{-12}m/V)$	$\lambda_v\ (\mu m)$	Refractive Index
LiNbO$_3$	$r_{13} = r_{23} = 8.6$	0.63	$n_o = 2.2967$
	$r_{33} = 30.8$		$n_e = 2.2082$
	$r_{22} = -r_{61} = -r_{12} = 3.4$		
	$r_{51} = r_{42} = 28$		
SiO$_2$	$r_{11} = -r_{21} = -r_{62} = 0.29$	0.63	$n_o = 1.546$
	$r_{41} = -r_{52} = 0.2$		$n_e = 1.555$
KDP	$r_{41} = r_{52} = 8.6$	0.55	$n_o = 1.50737$
(Potassium	$r_{63} = 10.6$		$n_e = 1.46685$
dihydrogen			
phosphate)			
ADP	$r_{41} = r_{52} = 2.8$	0.55	$n_o = 1.52$
(Ammonium	$r_{63} = 8.5$	0.55	$n_e = 1.48$
dihydrogen			
phosphate)			
GaAs	$r_{41} = r_{52} = r_{63} = 1.2$	0.9	$n_o = n_e = 3.42$
			(Cubic)

Example 1: Index ellipsoid of Lithium Niobate (LiNbO₃)

The linear electro-optic tensor, according to Table 5.2, is

$$[r_{ij}] = \begin{pmatrix} 0 & -r_{12} & r_{13} \\ 0 & r_{22} & r_{13} \\ 0 & 0 & r_{33} \\ 0 & r_{51} & 0 \\ r_{51} & 0 & 0 \\ -r_{22} & 0 & 0 \end{pmatrix}. \qquad (5.5\text{-}3)$$

For such a crystal and in the presence of an external electric field $\mathbf{E} = E_z\mathbf{a}_z$, the index ellipsoid equation can be reduced to

$$\frac{x^2}{n_o^2} + \frac{y^2}{n_o^2} + \frac{z^2}{n_e^2} + r_{13}E_z x^2 + r_{13}E_z y^2 + r_{33}E_z z^2 = 1,$$

which can be re-written as

$$\frac{x^2}{n_x^2} + \frac{y^2}{n_y^2} + \frac{z^2}{n_z^2} = 1, \tag{5.5-4a}$$

where

$$\frac{1}{n_x^2} = \frac{1}{n_o^2} + r_{13}E_z, \tag{5.5-4b}$$

$$\frac{1}{n_y^2} = \frac{1}{n_o^2} + r_{13}E_z, \tag{5.5-4c}$$

and

$$\frac{1}{n_z^2} = \frac{1}{n_e^2} + r_{33}E_z. \tag{5.5-4d}$$

Assuming $n_o^2 r_{13}E_z \ll 1$ and $n_e^2 r_{33}E_z \ll 1$ and with the help of the approximation that $\sqrt{1-\Delta} \approx 1-\Delta/2$ for $\Delta \ll 1$, we can find the refractive indices, from Eq. (5.5-4), for x, y, and z-polarized waves and they are

$$n_x = n_y \approx n_o - \frac{1}{2}r_{13}n_o^3 E_z \tag{5.5-5a}$$

and

$$n_z \approx n_e - \frac{1}{2}r_{33}n_e^3 E_z. \tag{5.5-5b}$$

Therefore, when the applied electrical field is along the z-direction, the ellipsoid does not undergo any rotation but only the lengths of the axes of the ellipsoid change. This deformation of the index ellipsoid creates the externally induced birefringence.

Example 2: Index ellipsoid of Potassium Dihydrogen Phosphate (KDP)

For a KDP crystal, its linear electro-optic tensor, according to Table 5.2, is

$$[r_{ij}] = \begin{pmatrix} 0 & 0 & 0 \\ 0 & 0 & 0 \\ 0 & 0 & 0 \\ r_{41} & 0 & 0 \\ 0 & r_{41} & 0 \\ 0 & 0 & r_{63} \end{pmatrix}. \tag{5.5-6}$$

In the presence of an external field $\mathbf{E} = E_x \mathbf{a}_x + E_y \mathbf{a}_y + E_z \mathbf{a}_z$ the index ellipsoid equation can be reduced to

$$\frac{x^2}{n_o^2} + \frac{y^2}{n_o^2} + \frac{z^2}{n_e^2} + 2r_{41}E_x yz + 2r_{41}E_y xz + 2r_{63}E_z xy = 1. $$

$$\tag{5.5-7}$$

The mixed terms in the equation of the index ellipsoid imply that the major and minor axes of the ellipsoid, with a field applied, are no longer parallel to the *x*, *y*, and *z* axes, which are the directions of the principal axes when no field is present. Indeed upon application of the external field, the index ellipsoid changes its orientation.

5.6 Some Applications of the Electro-Optic Effect

5.6.1 *Intensity modulation*

Longitudinal Configuration: A typical arrangement of an electro-optic *intensity modulator* is shown in Fig. 5.10. It consists of an electro-optic crystal placed between two crossed polarizers whose polarization axes are perpendicular to each other. The *polarization axis* defines the direction along which the emerging light is linearly polarized. We use a KDP crystal with its principal axes aligned with x, y, and z. An electric field is applied through the voltage V along the z-axis, which is the direction of propagation of the optical field, thus justifying the name *longitudinal configuration.* For E_z only, Eq. (5.5-7) then becomes

$$\frac{x^2}{n_o^2} + \frac{y^2}{n_o^2} + \frac{z^2}{n_e^2} + 2r_{63}E_z xy = 1. \qquad (5.6\text{-}1)$$

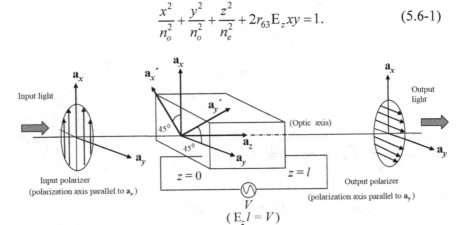

Fig. 5.10 Longitudinal electro-optic intensity modulation system.

We inspect Eq. (5.5-1) and see that there is one cross term xy present. From analytic geometry, we know that by rotating the x and y axes to a new x' and y' axes with respect to the z-axis, we can eliminate the cross term. Indeed, every quadratic expression

$$Ax^2 + Bxy + Cy^2 \qquad (5.6\text{-}2a)$$

can be reduced to the form

$$A'x'^2 + C'y'^2 \qquad (5.6\text{-}2b)$$

by rotating the axes through an angle θ where

$$\cot(2\theta) = (A - C) / B, \qquad\qquad (5.6\text{-}2c)$$

and

$$\begin{pmatrix} x \\ y \end{pmatrix} = \begin{pmatrix} \cos\theta & -\sin\theta \\ \sin\theta & \cos\theta \end{pmatrix} \begin{pmatrix} x' \\ y' \end{pmatrix}. \qquad\qquad (5.6\text{-}2d)$$

Now comparing Eq. (5-6-1) with Eq. (5.6-2a), we see that $A = C$, and hence $\cot 2\theta = 0$, which means $\theta = \pm 45°$. We take $\theta = -45°$, and Eq. (5.6-2d) becomes

$$x = \frac{1}{\sqrt{2}}(x' + y') \quad \text{and} \quad y = -\frac{1}{\sqrt{2}}(x' - y'). \qquad\qquad (5.6\text{-}3)$$

Introducing these expressions into Eq. (5.6-1), we have the index ellipsoid equation in the coordinate system aligned with the new principal axes, x', y' and z:

$$\left(\frac{1}{n_x'^2}\right)x'^2 + \left(\frac{1}{n_y'^2}\right)y'^2 + \frac{z^2}{n_e^2} = 1 \qquad\qquad (5.6\text{-}4)$$

where

$$\frac{1}{n_x'^2} = \frac{1}{n_o^2} - r_{63}E_z \quad \text{and} \quad \frac{1}{n_y'^2} = \frac{1}{n_o^2} + r_{63}E_z,$$

implying

$$n_x' \approx n_o + \frac{n_o^3}{2}r_{63}E_z \quad \text{and} \quad n_y' \approx n_o - \frac{n_o^3}{2}r_{63}E_z,$$

In Fig. 5.10, $\mathbf{a}_x'$ and $\mathbf{a}_y'$ are the unit vectors along the x'- and y'-axes, which have been rotated against the x- and y-axes by $45°$.

MATLAB Example: Index Ellipsoid

The index ellipsoid of a KDP crystal with external applied electric field along the principal z axis is given by Eq. (5.6-1). We shall plot this for a given applied electric field and show the ellipsoid rotates along the $x - y$ plane by $45°$. We start from Eq. (5.6-4) and for $z = 0$, it gives an ellipse on the $x' - y'$ plane and hence the y' values according to x' are calculated by:

$$y' = \pm n'_y \sqrt{1 - (1/n'_x)^2 x'^2} \ .$$

Since the y' value is defined in the real axis, the range of the x' is given by $-n'_x \le x' \le n'_x$. The refractive indices (n'_x, n'_y) along the x' and y' coordinates are given by Eq. (5.6-4). Using MATLAB, first we make a vector that represents the x' in the range $-n'_x \le x' \le n'$ and then we calculate the y' values according to the above equation. After that, we transform the calculated values (x', y') into (x, y) by

$$x = (1/\sqrt{2})(x' + y') \quad \text{and} \quad y = (-1/\sqrt{2})(x' - y') .$$

Using the 'plot(.)' function, we plot (x, y) for the figure, which is shown below for $E_z = 0.5 \times 10^{11}$ V/m. Note that this value of the applied field is unrealistic. We chose this value simply for the sake of illustration. The m-file used to generate Fig. 5.11 shown below is given in Table 5.3.

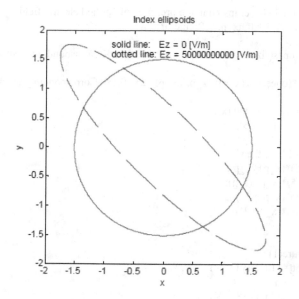

Fig. 5.11 Index Ellipsoid by MATLAB.

Table 5.3 Index_Ellipsoid.m (m-file for plotting Eq. (5.6-1)).

```
%Index_Ellipsoid.m
clear

no=1.51; %no is the refractive index of the ordinary axis of KDP
r63=10.6*10^-12; %Electro-optic Coefficient (r63)of KDP
E=input('Applied Electric Field [V/m]  = ') %suggested value is 0.5*10^ 11

for n=1:2


    if
    n==1
    Ez=0;
    end
    if
    n==2
    Ez=E;
    end
    nx_p=no+(no^3)/2*r63*Ez; %principal refractive index along x', see Eq. (5.6-4)
ny_p=no-(no^3)/2*r63*Ez; %principal refractive index along y'
```

```
%Calculate the index ellipsoid in the presence of applied electric field
x_p=-nx_p:nx_p/1000:nx_p; %Range of x'

y1_p=ny_p*(ones(size(x_p))-(x_p./nx_p).^2).^0.5; %Corresponding positive y'

y2_p=-ny_p*(ones(size(x_p))-(x_p./nx_p).^2).^0.5;%Corresponding negative y'

%Transform (x',y') to the (x,y)
%x11 and x12 are the x's

%y11 and y12 are the y's
x11=1/(2^0.5)*(x_p+y1_p);
y11=-1/(2^0.5)*(x_p-y1_p);
x12=1/(2^0.5)*(x_p+y2_p);
y12=-1/(2^0.5)*(x_p-y2_p);

  if n==1
          figure(1)
          plot(x11,real(y11),'-')
          hold on
          plot(x12,real(y12),'-')
   end

  if n==2
          plot(x11,real(y11),'--
          ')
          hold on
          plot(x12,real(y12),'--')
   end
end

text(-max(x11)*0.5, max(y11),['solid line:   Ez = ', num2str(0), ' [V/m]']);
text(-max(x11)*0.5, max(y11)*0.9,['dotted line: Ez = ', num2str(E), '
[V/m]']); axis square
hold off
xlabel('x')
ylabel('y')
title('Index ellipsoids')
```

Now, let us return to Fig. 5.10. The input light field at $z = 0$, just in front of the crystal, is polarized along the x direction and can be written as $E_0\mathbf{a}_x$. The field can be resolved into two mutually orthogonal components polarized along x' and y'. After passage through the electro-optic crystal for a distance l, the field components along $\mathbf{a}'_x$ and $\mathbf{a}'_y$ are

$$E_{x'}\big|_{z=l} = \frac{E_0}{\sqrt{2}}\exp(-j\frac{\omega_0}{c}n'_x l) \qquad (5.6\text{-}5a)$$

and

$$E_{y'}\big|_{z=l} = \frac{E_0}{\sqrt{2}}\exp(-j\frac{\omega_0}{c}n'_y l) \;, \qquad (5.6\text{-}5b)$$

respectively. The phase difference at $z = l$ between the two components is called the *retardation* Φ_L and is given by

$$\Phi_L = \frac{\omega_0}{c}(n'_x - n'_y)l$$

$$= \frac{\omega_0}{c}n_o^3 r_{63}V = \frac{2\pi}{\lambda_v}n_o^3 r_{63}V \;, \qquad (5.6\text{-}6)$$

where $V = E_z l$. At this point it is interesting to mention that at $\Phi_L = \pi$, the electro-optic crystal essentially acts as a half-wave plate where the birefringence is induced electrically. The crystal causes a x-polarized wave at $z = 0$ to acquire a y-polarized at $z = l$. The input light field then passes through the output polarizer unattenuated. With the electric field inside the crystal turned off ($V = 0$), there is no output light, as it is blocked off by the crossed output polarizer. The system, therefore, can switch light on and off electro-optically. The voltage yielding a retardation $\Phi_L = \pi$ is often referred to as the *half-wave voltage*,

$$V_\pi = \frac{\lambda_v}{2n_o^3 r_{63}} \;. \qquad (5.6\text{-}7)$$

From Table 5.2 and at $\lambda_v = 0.55\,\mu m$, $V_\pi \approx 7.58$ kV for KDP.

Returning to analyze the general system, the electric field component parallel to $\mathbf{a_y}$, that is, the component passed by the output polarizer, is

$$E_y^o = \left[E_{x'} \big|_{z=l} \mathbf{a'_x} + E_{y'} \big|_{z=l} \mathbf{a'_y} \right] \cdot \mathbf{a_y}$$

$$= \left[\frac{E_0}{\sqrt{2}} \exp(-j\frac{\omega_0}{c} n'_x l) \mathbf{a'_x} + \frac{E_0}{\sqrt{2}} \exp(-j\frac{\omega_0}{c} n'_y l) \mathbf{a'_y} \right] \cdot \mathbf{a_y}$$

Noting that $(\mathbf{a'_y}) \cdot \mathbf{a_y} = 1/\sqrt{2}$ and $(\mathbf{a'_x}) \cdot \mathbf{a_y} = -1/\sqrt{2}$ the above equation becomes

$$E_y^o = \frac{1}{2} E_0 \left[\exp\left(-j\frac{\omega_0}{c} n'_y l\right) - \exp\left(-j\frac{\omega_0}{c} n'_x l\right) \right].$$

Hence the ratio of the output $(I_0 = |E_y^o|^2)$ and the input $(I_i = |E_0|^2)$ intensities is

$$\frac{I_0}{I_i} = \sin^2\left[\frac{\omega_0}{2c} (n'_x - n'_y) l \right] = \sin^2\left(\frac{\Phi_L}{2} \right) = \sin^2\left(\frac{\pi}{2} \frac{V}{V_\pi} \right). \qquad (5.6\text{-}8)$$

Figure 5.12 shows a plot of the transmission factor (I_0 / I_i) versus the applied voltage V. Note that the most linear region of the curve is obtained for a bias voltage at $V_\pi / 2$. Therefore, the electro-optic modulator is usually biased with a fixed retardation to the 50 % transmission point. The bias can be achieved either electrically (by applying a fixed voltage $V = V_\pi / 2$) or optically (by using a quarter-wave plate). The quarter-wave plate has to be inserted between the electro-optic crystal and the output polarizer in such a way that its slow and fast axes are aligned in the direction of $\mathbf{a'_x}$ and $\mathbf{a'_y}$.

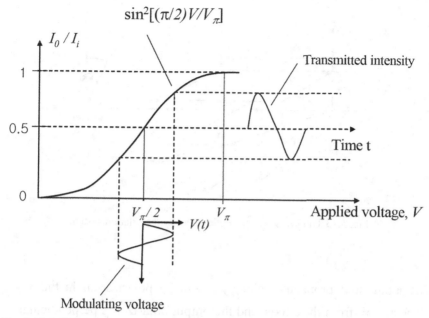

Fig. 5.12 Relationship between the modulating voltage and the transmitted intensity.

Transverse Configuration: In the previous discussion, the electric field was applied along the direction of light propagation. Consider, now, the configuration shown in Fig. 5.13, where the applied field is perpendicular to the direction of propagation. In this arrangement, the modulator operates in the *transverse configuration*.

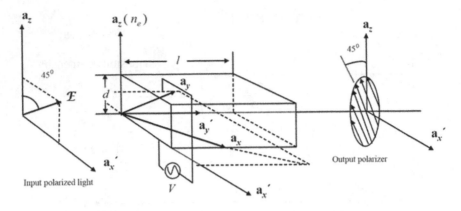

Fig. 5.13 Transverse electro-optic intensity modulation system.

The input light propagates along y', with its polarization in the x' - z plane at $45°$ from the z-axis, and the output polarizer is perpendicular to the incident polarization. Due to the applied voltage V, the principal axes x and y of the index ellipsoid are rotated to new principal axes along the x' and y' directions. Again, the change of the index ellipsoid produced by E_z is given by Eq. (5.6-4). Note that, in this configuration, there is no birefringence induced along the z-axis. Therefore, the z component of the incident electric field experiences no phase modulation. However, the $\mathbf{a}'_x$ component will experience retardation. Assuming that the input field polarization is $E_0(\mathbf{a}'_x + \mathbf{a}_z)/\sqrt{2}$, the field components $\mathbf{a}'_x$ and $\mathbf{a}_z$ at the exit of the crystals are then

$$E'_x = \frac{E_0}{\sqrt{2}} \exp\left(-j\frac{\omega_0}{c}n'_x l\right) \tag{5.6-9}$$

and

$$E_z = \frac{E_0}{\sqrt{2}} \exp\left(-j\frac{\omega_0}{c} n_e l\right), \tag{5.6-10}$$

respectively. The phase difference between the two components is

$$\Phi_T = \frac{\omega_0}{c}(n'_x - n_e)l$$

$$= \frac{\omega_0}{c}\left[(n_o - n_e) + \frac{n_o^3}{2} r_{63} E_z\right]l$$

$$= \frac{\omega_0}{c}\left[(n_o - n_e) + \frac{n_o^3}{2} r_{63} \frac{V}{d}\right]l, \tag{5.6-11}$$

where we have used the value of n'_x in Eq. (5.6-4), and d is the crystal length along the direction of the applied V. From Eq. (5.6-11), we can find the electrically induced half-wave voltage V_π for the transverse configuration by setting

$$\frac{\omega_0}{c}\frac{n_o^3}{2} r_{63} \frac{V_\pi}{d} l = \frac{2\pi}{\lambda_v}\frac{n_o^3}{2} r_{63} \frac{V_\pi}{d} l = \pi,$$

which gives the transverse half-wave voltage as

$$V_\pi = \left(\frac{\lambda_v}{n_o^3 r_{63}}\right)\frac{d}{l} = (V_\pi)_T,$$

By comparing this with Eq. (5.6-7) for the longitudinal half-wave voltage, $(V_\pi)_L$, we have

$$(V_\pi)_T = 2\frac{d}{l}(V_\pi)_L.$$

We can see that the transverse half-wave voltage is $2(d/l)$ times the longitudinal half-wave voltage. From Table 5.2 and at $\lambda_v = 0.55\mu m$, $d = 5mm$, $l = 10mm$, $V_\pi \approx 7.58$ kV for KDP in the transverse configuration. This value is the same as that of the half-wave voltage for KDP in the longitudinal configuration. However, we can obtain a lower half-wave voltage than that in the longitudinal configuration by employing a longer crystal along the propagation direction, whereas in the longitudinal configuration $(V_\pi)_L$ does not depend on l [see Eq. (5.6-7)]. Therefore, the transverse configuration is a more desirable mode of operation. Besides, in the transverse case the field electrodes do not interfere with the incident light beam.

Now, the final optical field passed by the output polarizer [along the $(\mathbf{a}_z - \mathbf{a}'_x)/\sqrt{2}$ direction] is

$$E_y^o = [E'_x \mathbf{a}'_x + E_z \mathbf{a}_z] \cdot [(\mathbf{a}_z - \mathbf{a}'_x)/\sqrt{2}]$$

$$= \frac{1}{\sqrt{2}}[E_z - E'_x]$$

$$= \frac{E_0}{\sqrt{2}}\left[\exp\left(-j\frac{\omega_0}{c}n_e l\right) - \exp\left(-j\frac{\omega_0}{c}n'_x l\right)\right], \qquad (5.6\text{-}12)$$

where we have used Eqs. (5.6-9) and (5.6-10). Similar to the longitudinal case, the ratio of the output ($I_0 = |E_y^o|^2$) to the input ($I_i = |E_0|^2$) intensities can be expressed in terms of the retardation as

$$\frac{I_0}{I_i} = \sin^2\left(\frac{\Phi_T}{2}\right). \qquad (5.6\text{-}13)$$

However, note that in the transverse case, the output field is, in general, elliptically polarized in the absence of the applied field because, from Eq. (5.6-11), $\Phi_T = (\omega_0/c)(n_o - n_e)l$ at $V = 0$, while $\Phi_L = 0$ at $V = 0$ for the longitudinal case. This provides a means for changing linearly polarized light to elliptically polarized light electro-optically in the transverse case.

5.6.2 *Phase modulation*

Figure 5.13 shows a system capable of phase-modulating the input light. Note that this system is similar to that of Fig. 5.10. In intensity modulation the input light is polarized along the bisector of the x' and y' axes, whereas in the *phase-modulation* scheme, the light is polarized parallel to one of the induced birefringence axes (x' in Fig. 5.14).

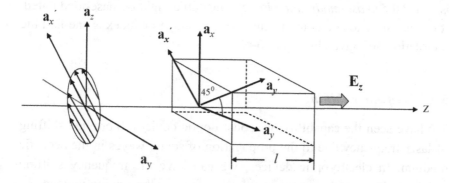

Fig. 5.14 An electro-optic phase modulating scheme.

The applied electric field along z direction does not change the state of polarization, but only the phase. The phase shift at the end of the crystal is

$$\phi'_x = \frac{\omega_0}{c} n'_x l, \qquad (5.6\text{-}14)$$

where n'_x is given by Eq. (5.6-4). As an example, if the applied field E_z is modulated according to

$$E_z = E_m \sin \omega_m t \qquad (5.6\text{-}15)$$

the light field at the exit of the crystal is

$$\mathcal{E} = \text{Re}[E'_x \exp(j\omega_0 t)]\mathbf{a}'_x = \text{Re}\{E_0 \exp(j(\omega_0 t - \phi'_x))\}\mathbf{a}'_x$$

$$= \text{Re}\left\{E_0 \exp(j\omega_0 t)\exp\left(\frac{-j\omega_0 n_0}{c}\right)\exp(-j\delta \sin \omega_m t)\right\}\mathbf{a}'_x,$$

where

$$\delta = \frac{\omega_0 n_o^3 r_{63} E_m l}{2c} = \frac{\pi n_o^3 r_{63} E_m l}{\lambda_v}$$

is called the *phase-modulation index*. The output light is phase-modulated. It is interesting to point out that the phase-modulation index is one-half the retardation Φ_L given by Eq. (5.6-6).

5.6.3 *Frequency shifting*

We have seen the capability of acousto-optic devices to provide shifting of laser frequency due to the propagation of sound waves in the acoustic medium. In electro-optic devices, we can have the frequency shifting capability by providing a saw-tooth waveform to the driving voltage. As an example, we shall use the transverse configuration shown in Fig. 5.13 to explain frequency shifting. We assume the input light is now polarized along the x' direction. In addition, the output polarizer is removed. Under the situation, the electric field at the exit of the crystal can be expressed as

$$\mathbf{E} = \mathrm{Re}[E'_x \exp(j\omega_0 t)\exp(-j\phi'_x)]\mathbf{a}'_\mathbf{x}, \qquad (5.6\text{-}16)$$

where again $\phi'_x = \dfrac{\omega_0}{c}n'_x l$ and $n'_x \approx n_o + \dfrac{n_o^3}{2}r_{63}E_z$ as given by eq. (5.6-4) for KDP. Therefore,

$$\phi'_x = \frac{\omega_0}{c}n'_x l = \frac{\omega_0}{c}\left(n_0 + \frac{n_0^3}{2}r_{63}E_z\right)l$$

$$= \frac{\omega_0}{c}n_0 l + \frac{n_0^3}{2}r_{63}E_z l$$

$$= \frac{2\pi}{\lambda_v}n_0 l + \frac{\pi}{\lambda_v}n_0^3 r_{63}\frac{V(t)}{d}l$$

where $V(t)$ is the driving voltage. Using the definition of the half-wave voltage, $(V_\pi)_T = \left(\dfrac{\lambda_v}{n_0^3 r_{63}}\right)\dfrac{d}{l}$. ϕ'_x can be expressed as

$$\phi'_x = \frac{2\pi}{\lambda_v} n_0 l + \frac{\pi V(t)}{(V_\pi)_T}.$$

Since the instantaneous frequency is $\Delta\omega = \dfrac{d\phi'_x}{dt}$, we then have

$$\Delta\omega = \frac{d\phi'_x}{dt} = \frac{\pi}{(V_\pi)_T} \frac{dV(t)}{dt} \tag{5.6-17}$$

Equation (5.6-16) then can be written as

$$\begin{aligned}
\boldsymbol{E} &= \mathrm{Re}[E_0 \exp(j\omega_0 t)\exp(-j\phi'_x)]\mathbf{a}'_x \\
&= \mathrm{Re}[E_0 \exp(j\omega_0 t)\exp(-j\Delta\omega t)]\mathbf{a}'_x
\end{aligned} \tag{5.6-18}$$

to reflect a frequency change of the field. Let us now find the frequency, $\Delta\omega$, for a given $V(t)$ shown in the Fig. 5.15. The driving voltage is a saw-tooth waveform given by

$$V(t) = \frac{2V_a}{T} t \quad \text{volt} \quad \text{for} \quad -T/2 < t < T/2.$$

For this waveform, according to Eq. (5.6-17),

$$\Delta\omega = \frac{2\pi V_a}{(V_\pi)_T T} = \frac{2\pi}{T} \quad \text{for} \quad V_a = (V_\pi)_T.$$

Hence Eq. (5.6-18) becomes

$$\boldsymbol{E} = \mathrm{Re}[E_0 \exp[j(\omega_0 - \Delta\omega)t]]\mathbf{a}'_x$$

with $\Delta\omega = 2\pi/T$ and the down-shifting in the frequency is controlled by the period T of the saw-tooth waveform.

For heterodyning using electro-optic devices, the input light propagates along y', with its polarization in the $x'-z$ plane at 45° from the z axis as shown in Fig. 5.13. Under this situation, the electric field along the $\mathbf{a}'_x$ direction will acquire frequency shifting upon exiting from the crystal as we have just described, while the frequency of electric field along the $\mathbf{a}_z$ direction will remain the same as that of the input electric field. The two polarizations carrying different temporal frequencies can be separated, for example, using a beamsplitter at the exit of the crystal, followed by a pair of polarizers oriented perpendicular to each other. Electro-optic heterodyning in holography has been demonstrated in some recent publications [Liu *et al.* (2016), Kim *et al.* (2016)].

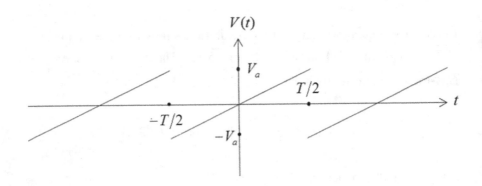

Fig. 5.15 Saw-tooth waveform for the driving voltage.

Problems:

5.1 For a time-harmonic uniform plane wave in a linear isotropic homogeneous medium, we assume that the fields vary according to $\exp[j(\omega_0 t - \mathbf{k}_0 \cdot \mathbf{R})]$. Show that in this case, Maxwell's equations in a source-free region can be expressed as

$$\mathbf{k}_0 \cdot \boldsymbol{\mathcal{D}} = 0,$$

$$\mathbf{k}_0 \cdot \boldsymbol{\mathcal{H}} = 0,$$

$$\mathbf{k}_0 \times \boldsymbol{\mathcal{E}} = \omega_0 \mu \boldsymbol{\mathcal{H}},$$

$$\mathbf{k}_0 \times \boldsymbol{\mathcal{H}} = -\omega_0 \boldsymbol{\mathcal{D}},$$

by first verifying Eq. (5.2-1). Subsequently, verify Eq. (5.2-3).

5.2 Show that, for plane wave propagation in anisotropic media, the phase velocity is $u_p = |\mathbf{u_g}| \cos\alpha$, where α is the angle between the direction of propagation of the wavefront and $S = \boldsymbol{\mathcal{E}} \times \boldsymbol{\mathcal{H}}$, and $|\mathbf{u_g}|$ is the velocity of the energy propagation

5.3 Given a beam of light that might be either unpolarized or circularly polarized, how might you determine its actual state of polarization?

5.4 Starting from Eq. (5.2-14), where the total phase velocity of a plane wave propagating at an angle θ with respect to the optic axis as illustrated in Fig. 5.3 is given by

$$u = \sqrt{(u_2 \cos\theta)^2 + (u_3 \sin\theta)^2},$$

derive Eq. (5.4-5).

5.5   Starting from Eq. (5.2-18), verify Eq. (5.2-19).

5.6   For lithium niobate as the uniaxial crystal in Fig. 5.4 and for propagation at an angle of $45°$ from the optic axis, find the angle between $\mathbf{E}$ and $\mathbf{D}$ associated with the o-ray and the e-ray.

5.7   A clockwise circularly polarized light ($\lambda_v = 0.688\ \mu m$) described by

$$\mathbf{E}_i = E_i \cos(\omega_0 t - k_0 x)\mathbf{a_y} - E_i \sin(\omega_0 t - k_0 x)\mathbf{a_z}$$

is incident on an x-cut calcite crystal ($n_o = 1.658, n_e = 1.486$) of thickness $d = 0.006$mm. Find the expression of the emergent field $\mathbf{E}_{out}$ and its state of polarization.

5.8   A clockwise circularly polarized light ($\lambda_v = 0.688 \mu m$) is incident on an x-cut quartz crystal ($n_o = 1.544, n_e = 1.553$) of thickness $d = 0.025$ mm. Determine the state of polarization of the emerging beam.

5.9   Consider a Lithium Niobate crystal, where we apply an external field along the y-principal direction, i.e., $\mathbf{E} = E_y \mathbf{a_y}$,

   a) In what plane the ellipsoid is rotated through an angle?
   b) Give an expression for the rotated angle, assuming small angles, i.e., $\sin\theta \approx \theta$ and $\cos\theta \approx 1$. Also find the rotated angle for a voltage of 1 KV across a 1-mm thick crystal.

5.10   Consider a KDP crystal, where we apply an external field along the x-principal direction, E_x.

   a) Show that the ellipsoid is rotated through an angle α in the yz-plane, where the angle is defined by

$$\tan 2\alpha = -\frac{2r_{41}E_x}{(1/n_e^2 - 1/n_o^2)}.$$

b) For an applied electric field of 10^{6} V/m, find the angle of rotation.

c) Show that, for small angle assumptions, the principal indices of refraction are approximately given by the following expressions:

$$n'_y = n_o + \frac{n_o^5 (n_e r_{41} E_x)^2}{2(n_o^2 - n_e^2)},$$

$$n'_z = n_e - \frac{n_e^5 (n_o r_{41} E_x)^2}{2(n_o^2 - n_e^2)}.$$

5.11   In the absence of an applied electric field, the ellipsoid for the uniaxial crystal LiNbO₃ is given by Eq. (5.4-1). An external electric field E_{z0} is now applied along the optic axis (z) of the crystal.

(a) Find the equation of the index ellipsoid and estimate the corresponding refractive indices along the x, y, and z directions.

(b) Assume that linearly polarized light is incident along the y axis, and the incident wave is decomposed into E_x and E_z components. Find the phase retardation between the two components after the light traverses the length of the crystal, l.

(c) Calculate the required magnitude of E_{z0} if the emerging wave is circularly polarized (Take $l = 1$ cm).

Bibliography

Banerjee, P.P. and T.-C. Poon (1991). *Principles of Applied Optics.* Irwin, Homewood, Ill

Chen, Chin-Lin (1996). *Elements of Optoelectronics & Fiber Optics.* Irwin,
Chicago, Illinois.

Fowles, G. R. (1975). *Introduction to Modern Optics.* Holt, Rinehart and Winston, New Y

Ghatak, A.K. and K. Thyagarajan (1989). *Optical Electronics.* Cambridge Univer
Cambridge.

Haus, H.A. (1984). *Waves and Fields in Optoelectronics.* Prentice-Hall, Inc. New Jersey.

Johnson, C.C. (1965). *Field and Wave Electrodynamics.* McGraw-Hill, New York.

Kim, H. *et al.* (2016). "Full-Color Optical Scanning Holography with Common Red, Greer
Channels [Invited]," Applied Optics, 55, pp. A17- A-20.

Liu, J.-P. *et al.* (2016). "Coherence Experiments in Single-Pixel Digital Holography," Opt
40, pp. 2366- 2369.

Nussbaum, A and R.A. Phillips (1976). *Contemporary Optics for Scientists and Engineer.*
Hall, Inc. New Jersey.

Yariv, A. (1976). *Introduction to Optical Electronics.* 2nd Ed. Holt, Rinehart, and Wir
York.

Index

A

ABCD matrix, 25,42,129
aberration, 24,
absorption, 179, 196, 259
acoustic field, 236
acoustic frequency, 188
acoustic medium, 188–190, 194,
196, 199, 205, 211, 212, 214, 248,
294
acoustic radiation, 214, 251
acoustic spectrum, 226
acoustic transducer, 223, 228
acoustic wave, 228, 252
acousto-optic beam, 236
acousto-optic bistable device, 228
acousto-optic Bragg cell, 222
acousto-optic Bragg diffraction,
248, 251
acousto-optic cell, 204–205, 229,
250
acousto-optic diffraction, 214, 234,
239, 252
acousto-optic effect, 198, 222, 226
acousto-optic FM demodulator, 228
acousto-optic heterodyne, 246
acousto-optic interaction, 188, 194,
234
acousto-optic modulator, 188, 197,
211, 214, 225, 232, 234, 239, 248,
251–253
acousto-optic spatial filtering, 234
acousto-optics, 62, 138, 188, 200,
204–205, 230, 234, 251–252, 276
ADP, 279
Ampere, 61
amplification, 196, 251
amplifier, 229–230
amplitude reflection and

transmission coefficients, 78, 135
analytic geometry, 282
analyzer, 190, 225–226, 249
angular frequency, 66
angular magnification, 27
angular plane-wave spectrum, 186
angular spectrum, 196, 214
anisotropic, 62, 71, 239, 254–255,
261, 276, 297
anisotropic binding, 254–255
anisotropic crystal, 261
anisotropic media, 62, 71, 254, 261,
276, 297
anisotropy, 254, 262
antenna, 248
aperture, 102, 104–105, 107, 116,
119, 196–197
atmosphere, 102, 104–105, 107,
116, 119, 196–197
attenuation, 196
average, 196
average power, 83

B

back focal plane, 27, 107–109, 248
back focal point, 27, 38, 42
back focus, 34–35, 130
Baker-Hausdorff formula, 154, 187
Bandwidth, 188
beam profile, 113, 131, 143, 158,
161, 164, 167, 180, 183, 242
beam propagation, 59, 127, 142,
150, 153, 156–157, 162, 165, 168,
beam propagation method (BPM),
154, 156–157
beam waist, 150, 158, 161,164,
180, 184
Bessel function, 67, 207, 217

Printed in the United States
By Bookmasters